"At once deeply persuasive and deeply unsettling."
—Elizabeth Kolbert, Pulitzer Prize–winning author of *The Sixth Extinction*

THE WATER WILL COME

Rising Seas, Sinking Cities, and the Remaking of the Civilized World

JEFF GOODELL

Praise for Jeff Goodell's
THE WATER WILL COME

One of the Best Books of the Year

*New York Times * Washington Post * Booklist*

"Once you've read an excellent book about climate change, which Jeff Goodell's *The Water Will Come* most certainly is, you can never unremember the facts...Goodell has been writing about climate change for many years...He's the real deal, committed and making house calls."

—Jennifer Senior, *New York Times*

"If there was ever a moment when Americans might focus on drainage, this is it. But this fine volume concentrates on the slower and more relentless toll that water will take on our cities and our psyches in the years to come."

—Bill McKibben, *Washington Post*

"For people who want to learn more about climate change, rising sea levels and what it means for our future, read *The Water Will Come.*"

—Chris Hayes, *MSNBC*

"One of the most important books of the year...A potent examination not of whether seas will rise in our lifetimes, but of the fact that they will rise."

—The Sierra Club

"Goodell dives into a wonky, but vivid, mix of science, history, and sociology...Goodell talks about climate change and what it means to every person on the planet in a way that will engage even the non-*Nova* crowd."

—Zlati Meyer, *USA Today*

"Jeff Goodell's latest contribution to the environmental cause paints an eye-opening portrait of humankind's dilemma as temperatures—and sea levels—continue to rise. *The Water Will Come* brings together compelling anecdotes from all over the globe and shocking expert assessments that should make the world's few remaining skeptics reconsider. Read this book for a reminder of the stakes—right now, today—and why we have to work harder, faster, to address the climate challenge."

—John F. Kerry, former secretary of state

"*The Water Will Come* is absolutely brilliant scientific journalism, and certainly is a must-read for all of the world's citizens."

—*Forbes*

"This harrowing, compulsively readable, and carefully researched book lays out in clear-eyed detail what Earth's changing climate means for us today, and what it will mean for future generations…It's a thriller in which the hero in peril is *us*."

—John Green

"Engaging…Goodell points out that while sea levels have always risen and fallen, the current rise is driven primarily by the dramatically accelerating melting of the arctic ice caps, and with so many cities on seashores, this will be devastating."

—Dan Kaplan, *Booklist*

"Other writers have told the story of sea-level rise, but perhaps none as compellingly as Goodell. His riveting stories…clarify the implications of sea-level rise and the choices communities face."

—Robert Glennon, *New York Times Book Review*

"Goodell's journalistic writing style is engaging and will be accessible to a wide audience... This thought-provoking tour through our watery futures offers both challenge and inspiration."
—Jessica Lamond, *Science*

"Jeff Goodell has taken on some of the most important issues of our time, from coal mining to geoengineering. In *The Water Will Come*, he explains the threat of sea level rise with characteristic rigor and intelligence. The result is at once deeply persuasive and deeply unsettling."
—Elizabeth Kolbert, Pulitzer Prize–winning and *New York Times* bestselling author of *The Sixth Extinction*

"A must-read... Goodell writes with insight and compassion, giving us a primer we can use to persuade neighbors, friends, and politicians to take action now."
—The Leonardo DiCaprio Foundation

"Sea-level rise is coming. We know this as clearly as we know thermometer measurements, the melting point of ice, and the law of thermal expansion. Jeff Goodell's book cuts through the fossil-fuel lies and is a warning I hope we heed while there's still time."
—Senator Sheldon Whitehouse

"*The Water Will Come* is a well-rounded, persuasive survey... A frightening, scientifically grounded, and starkly relevant look at how climate change will affect coastal cities."
—*Kirkus Reviews*

"Scrupulously researched yet written in the clean and accessible style of a journalist who's perfected his craft... Persuasive, timely, and vividly constructed, *The Water Will Come* might be one of the most essential reads of the year."
—Amy Brady, *Shelf Awareness*

THE WATER WILL COME

Rising Seas, Sinking Cities,
and the Remaking
of the Civilized World

JEFF GOODELL

Back Bay Books
Little, Brown and Company
New York   Boston   London

Back Bay Books / Little, Brown and Company
Hachette Book Group
1290 Avenue of the Americas, New York, NY 10104
littlebrown.com

Originally published in hardcover by Little, Brown and Company, October 2017
First Back Bay paperback edition, August 2018

Back Bay Books is an imprint of Little, Brown and Company, a division of Hachette Book Group, Inc. The Back Bay Books name and logo are trademarks of Hachette Book Group, Inc.

The publisher is not responsible for websites (or their content) that are not owned by the publisher.

The Hachette Speakers Bureau provides a wide range of authors for speaking events. To find out more, go to hachettespeakersbureau.com or call (866) 376-6591.

ISBN 978-0-316-26024-4 (hc) / 978-0-316-26020-6 (pb)
Library of Congress Control Number: 2017947000

10 9 8 7 6 5 4

LSC-C

Printed in the United States of America

For Milo, Georgia, and Grace

Contents

THE WATER WILL COME

There's always that moment in a country's history when it becomes obvious the earth is less manageable than previously thought.

—Jim Shepard, "The Netherlands
Live with Water"

PROLOGUE: ATLANTIS

AFTER THE HURRICANE hit Miami in 2037, a foot of sand covered the famous bow-tie floor in the lobby of the Fontainebleau Hotel in Miami Beach. A dead manatee floated in the pool where Elvis had once swum. Most of the damage came not from the hurricane's 175-mile-an-hour winds, but from the twenty-foot storm surge that overwhelmed the low-lying city. In South Beach, historic Art Deco buildings were swept off their foundations. Mansions on Star Island were flooded up to their cut-glass doorknobs. A seventeen-mile stretch of Highway A1A that ran along the famous beaches up to Fort Lauderdale disappeared into the Atlantic. The storm knocked out the wastewater-treatment plant on Virginia Key, forcing the city to dump hundreds of millions of gallons of raw sewage into Biscayne Bay. Tampons and condoms littered the beaches, and the stench of human excrement stoked fears of cholera. More than three hundred people died, many of them swept away by the surging waters that submerged much of Miami

Beach and Fort Lauderdale; thirteen people were killed in traffic accidents as they scrambled to escape the city after the news spread—falsely, it turned out—that one of the nuclear reactors at Turkey Point, an aging power plant twenty-four miles south of Miami, had been heavily damaged by the surge and had sent a radioactive cloud floating over the city.

The president, of course, said that Miami would be back, that Americans did not give up, that the city would be rebuilt better and stronger than it had been before. But it was clear to those not fooling themselves that this storm was the beginning of the end of Miami as a booming twenty-first-century city.

All big hurricanes are disastrous. But this one was unexpectedly bad. With sea levels more than a foot higher than they'd been at the dawn of the century, much of South Florida was wet and vulnerable even before the storm hit. Because of the higher water, the storm surge pushed deeper into the region than anyone had imagined it could, flowing up drainage canals and flooding homes and strip malls several miles from the coast. Despite newly elevated runways, Miami International Airport was shut down for ten days. Salt water shorted out underground electrical wiring, leaving parts of Miami-Dade County dark for weeks and contaminated municipal drinking-water wells, leaving thousands of displaced people scrambling for bottled water that was air-dropped by the National Guard. In soggy neighborhoods, mosquitoes carrying Zika and dengue fever viruses hatched (injecting male mosquitoes with the *Wolbachia* bacteria, which public health officials had once hoped would inhibit the mosquitoes' ability to transmit the viruses, had failed when the *Aedes aegypti* mosquitoes that carry the diseases developed immunity to the bacteria). In Homestead, a low-lying working-class city in southern Miami-Dade County that had been flattened by Hurricane

Andrew in 1992, thousands of abandoned homes were bull-dozed because they were deemed a public health hazard. In Miami Shores, developers approached city officials with pro-posals to buy out entire blocks of waterlogged apartments, then dredge the streets and turn them into canals lined with houseboats. But financing for these projects never materialized.

Before the storm hit, damage from rising seas had already pushed city and county budgets to the brink. State and fed-eral money was scarce too, in part because Miami was seen by many Americans as a rich, self-indulgent city that had ignored decades of warnings about building too close to the water. Attempts had been made to armor the shore with sea-walls and elevate buildings, but only a small percentage of the richest property owners took protective action. The beaches were mostly gone too. The Feds decided they couldn't afford to spend $100 million every few years to pump in fresh sand, and without replenishment, the ever-higher tides car-ried the beaches away. By the late 2020s, the only beaches that remained were privately maintained oases of sand in front of expensive hotels. The hurricane took care of those, leaving the hotels and condo towers perched on limestone crags. Tourists disappeared. After the hurricane, the city became a mecca for slumlords, spiritual healers, and lawyers. In the parts of the county that were still inhabitable, only the wealthiest could afford to insure their homes. Mortgages were nearly impossible to get, mostly because banks didn't believe the homes would be there in thirty years.

Still, the waters kept rising, nearly a foot each decade. Each big storm devoured more of the coastline, pushing the water deeper and deeper into the city. The skyscrapers that had gone up during the boom years were gradually abandoned and used as staging grounds for drug runners and exotic-animal traffickers. Crocodiles nested in the ruins of the Frost Museum of Science. (Historians dryly noted that the namesake of the museum, billionaire Phillip Frost, had been a climate change skeptic.) Still, the waters kept rising. By the end of the twenty-first century, Miami became something else entirely: a popular diving spot where people could swim among sharks and barnacled SUVs and explore the wreckage of a great American city.

That is, of course, merely one possible vision of the future. There are brighter ways to imagine it—and darker ways. But I am a journalist, not a Hollywood screenwriter. In this book, I want to tell a true story about the future we are creating for ourselves, our children, and our grandchildren. It begins with this: the climate is warming, the world's great ice sheets are melting, and the water is rising. This is not a speculative idea, or the hypothesis of a few wacky scientists, or a hoax perpetrated by the Chinese. Sea-level rise is one of the central facts of our time, as real as gravity. It will reshape our world in ways most of us can only dimly imagine.

My own interest in this story began with an actual hurricane. Shortly after Hurricane Sandy hit New York City in 2012, I visited the Lower East Side of Manhattan, one of the neighborhoods that had been hardest hit by flooding from the storm. The water had receded by the time I arrived, but the neighborhood already smelled of mold and rot. The power was out, the shops were closed. I saw broken trees, abandoned cars, debris scattered everywhere, people hauling ruined furniture out of basement apartments. Dark waterlines were visible on many shop windows and doors. The surge in the East River had been more than nine feet high, overwhelming the seawall and inundating the low-lying parts of Lower Manhattan. As I walked around, watching people slowly put their lives back together, I wondered what would have happened if, instead of flooding the city and then receding in a few hours, the Atlantic Ocean had come in and stayed in.

I have been writing about climate change for more than a decade, but seeing the flooding on the Lower East Side made it visceral for me (I hadn't visited New Orleans until several years after Katrina hit—the TV images of the flooding there, catastrophic as they were, did not affect me as strongly as my walk through the Lower East Side). A year or so before Sandy rolled in, I had interviewed NASA scientist James Hansen, the godfather of climate change science, who told me that if nothing was done to slow the burning of fossil fuels, sea levels could be as much as ten feet higher by the end of the century. At the time, I didn't grasp the full implications of this. After Sandy, I did.

Soon after my visit to Lower Manhattan, I found myself in Miami, learning about the porous limestone foundation the city is built on and the flatness of the topography. During high tide, I waded knee-deep through dark ocean water in several Miami Beach neighborhoods; I saw high water backing up into working-class neighborhoods far to the west, near the border of the Everglades. It didn't take a lot of imagination to see that I was standing in a modern-day Atlantis-in-the-making. It became clear to me just how poorly our world is prepared to deal with the rising waters. Unlike, say, a global pandemic, sea-level rise is not a direct threat to human survival. Early humans had no problem adapting to rising seas— they just moved to higher ground. But in the modern world, that's not so easy. There's a terrible irony in the fact that it's the very infrastructure of the Fossil Fuel Age—the housing and office developments on the coasts, the roads, the railroads, the tunnels, the airports—that makes us most vulnerable.

*   *   *

Rising and falling seas represent one of the ancient rhythms of Earth, the background track that has played during the entire four-billion-year life of the planet. Scientists have understood this for a long time. Even in relatively recent history, sea levels have fluctuated wildly, driven by wobbles in the Earth's orbit that change the angle and intensity of the sunlight hitting the planet and cause ice ages to come and go. One hundred and twenty thousand years ago, during the last interglacial period, when the temperature of the Earth was very much like it is today, sea levels were twenty to thirty feet higher. Then, twenty thousand years ago, during the peak of the last ice age, sea levels were four hundred feet lower.

What's different today is that humans are interfering with this natural rhythm by heating up the planet and melting the vast ice sheets of Greenland and Antarctica. Until just a few decades ago, most scientists believed these ice sheets were so big and so indomitable that not even seven billion humans with all their fossil-fuel-burning toys could have much impact on them in the short term. Now they know better.

In the twentieth century, the oceans rose about six inches. But that was before the heat from burning fossil fuels had much impact on Greenland and Antarctica (about half of the recorded sea-level rise in the twentieth century came from the expansion of the warming oceans). Today, seas are rising at more than twice the rate they did in the last century. As warming of the Earth increases and the ice sheets begin to feel the heat, the rate of sea-level rise is likely to increase rapidly. A 2017 report by the National Oceanic and Atmospheric Administration, the United States' top climate science agency, says global sea-level rise could range from about one foot to more than eight feet by 2100. Depending on how much we heat up the planet, it will continue rising for centuries after that. Although there is still some uncertainty about these forecasts, many scientists I've talked to now believe that the high-end projections are likely to increase as they get a better understanding of ice dynamics. Temperature-wise, the trend lines are rising: 2016 was the hottest year on record (2017 was a close runner up), and as I'm writing this, the Arctic is thirty-six degrees warmer than normal.

But if you live on the coast, what matters more than the height the seas rise to is the *rate* at which they rise. If the water rises slowly, it's not such a big deal. People will have time to elevate roads and buildings and build seawalls. Or move away. It is likely to be disruptive but manageable. Unfortunately, Mother Nature is not always so docile. In the past, the seas have risen in dramatic pulses that coincide with the sudden collapse of ice sheets. After the end of the last ice age, there is evidence that the water rose about thirteen feet in a single century. If that were to occur again, it would be a catastrophe for coastal cities around the world, causing hundreds of millions of people to flee from the coastlines and submerging trillions of dollars' worth of real estate and infrastructure.

The best way to save coastal cities is to quit burning fossil fuels (if you're still questioning the link between human activity and climate change, you're reading the wrong book). But even if we ban coal, gas, and oil tomorrow, we won't be able to turn down the Earth's thermostat immediately. For one thing, carbon dioxide (CO_2) is not like other kinds of air pollution, such as the chemicals that cause smog, which go away as soon as you stop dumping them into the sky (as happened, by and large, when catalytic converters were installed on cars). A good fraction of the CO_2 emitted today will stay in the atmosphere for thousands of years. That means that even if we did reduce CO_2 tomorrow, we can't shut off the warming from the CO_2 we've already dumped into the air. "The climatic impacts of releasing fossil fuel CO_2 to the atmosphere will last longer than Stonehenge," scientist David Archer writes. "Longer than time capsules, longer than nuclear waste, far longer than the age of human civilization so far."

For sea-level rise, the slow response of the Earth's climate system has enormous long-term implications. Even if we replaced every SUV on the planet with a skateboard and every coal plant with a solar panel and could magically reduce global carbon pollution to zero by tomorrow, because of the heat that has already built up in the atmosphere and the oceans, the seas would not stop rising—at least until the Earth cooled off, which could take centuries.

This doesn't mean that cutting CO_2 is pointless. On the contrary. If we can hold the warming to about three degrees Fahrenheit above preindustrial temperatures, we might only face two feet of sea-level rise this century, giving people more time to adapt. However, if we don't end the fossil fuel party, we're headed for more than eight degrees Fahrenheit of warming—and with that, all bets are off. We could get four feet of sea-level rise by the end of the century—or we could get thirteen feet. The long-term consequences are even more alarming. If we burn all the known reserves of coal, oil, and gas on the planet, seas will likely rise by more than two hundred feet in the coming centuries, submerging virtually every major coastal city in the world.

The tricky thing about dealing with sea-level rise is that it's impossible to witness by just hanging out at the beach for a few weeks. Instead, the rise will make itself felt in higher storm surges, higher tides, and a gradual washing away of beaches, of roads, of coastal infrastructure. Even in the worst-case scenarios, the changes will occur over years and decades and centuries, not seconds and minutes and hours. It's exactly the kind of threat that we humans are genetically ill equipped to deal with. We have evolved to defend ourselves from a guy with a knife or an animal with big teeth, but we are not wired to make decisions about barely perceptible threats that gradually accelerate over time. We're not so different from the proverbial frog that boils to death in a pot of slowly warming water.

One architect I met while researching this book joked that with enough money, you can engineer your way out of any-thing. I suppose it's true. If you had enough money, you could raise or rebuild every street and building in Miami by ten feet and the city would be in pretty good shape for the next century or so. But we do not live in a world where money is no object, and one of the hard truths about sea-level rise is that rich cities and nations can afford to build seawalls, upgrade sewage systems, and elevate critical infrastructure. Poor cities and nations cannot. But even for rich countries, the economic losses will be high. One recent study estimated that with six feet of sea-level rise, nearly $1 trillion worth of real estate in the United States will be underwater, including one in eight homes in Florida. If no significant action is taken, global damages from sea-level rise could reach $100 trillion a year by 2100.

But it is not just money that will be lost. Also gone will be the beach where you first kissed your boyfriend; the man-grove forests in Bangladesh where Bengal tigers thrive; the crocodile nests in Florida Bay; Facebook headquarters in Sil-icon Valley; St. Mark's Basilica in Venice; Fort Sumter in Charleston, South Carolina; America's biggest naval base in Norfolk, Virginia; NASA's Kennedy Space Center; graves on the Isle of the Dead in Tasmania; the slums of Jakarta, Indo-nesia; entire nations like the Maldives and the Marshall

Islands; and, in the not-so-distant future, Mar-a-Lago, the winter White House of President Donald Trump. Globally, about 145 million people live three feet or less above the current sea level. As the waters rise, millions of these people will be displaced, many of them in poor countries, creating generations of climate refugees that will make today's Syrian war refugee crisis look like a high school drama production.

The real x factor here is not the vagaries of climate science, but the complexity of human psychology. At what point will we take dramatic action to cut CO_2 pollution? Will we spend billions on adaptive infrastructure to prepare cities for rising waters—or will we do nothing until it is too late? Will we welcome people who flee submerged coastlines and sinking islands—or will we imprison them? No one knows how our economic and political system will deal with these challenges. The simple truth is, human beings have become a geological force on the planet, with the power to reshape the boundaries of the world in ways we didn't intend and don't entirely understand. Every day, little by little, the water is rising, washing away beaches, eroding coastlines, pushing into homes and shops and places of worship. As our world floods, it is likely to cause immense suffering and devastation. It is also likely to bring people together and inspire creativity and camaraderie in ways that no one can foresee. Either way, the water is coming. As Hal Wanless, a geologist at the University of Miami, told me in his deep Old Testament voice as we drove toward the beach one day, "If you're not building a boat, then you don't understand what's happening here."

1. THE OLDEST STORY
EVER TOLD

THE R/V *KNORR* was a storied ship in the annals of science, known for its ability to take a pounding in rough seas and its unusual arrangement of propellers in the bow and stern that made it highly maneuverable. Scientists had used the *Knorr*, a 244-foot steel-hulled research vessel that was operated by Woods Hole Oceanographic Institution, for thousands of research expeditions around the world, including one that led to the discovery of the wreck of the RMS *Titanic*. A few years ago, I spent a month aboard the *Knorr*. On my trip, we were looking for nothing more glamorous than good, thick mud on the floor of the North Atlantic. By drilling cores in the mud and analyzing the shells of creatures buried within it, researchers can better understand past ocean temperature and salinity, which are important as scientists attempt to reconstruct the history of the Earth's climate.

Most of our time was spent cruising around the Bermuda

Rise, a cluster of extinct underwater volcanos near Bermuda, pausing to take core samples of the mud whenever the conditions looked good. At one point, we were about a hundred miles off the coast of New York City when we drifted over a place known to scientists as the Hudson Canyon, which is where the Hudson River used to drop over the continental shelf twenty thousand years ago when seas were lower. From the ship, a device called an acoustic echo machine painted colorful real-time images of the canyon as we passed over it. It was a remarkable sight: you could see where the Hudson had cut a path through the shelf, creating terraces and high walls. The canyon extends over 450 miles across the shelf, eventually reaching a depth of over 10,000 feet. "It's bigger than the Grand Canyon," Lloyd Keigwin, the chief scientist on the trip, told me as I looked down in wonder.

Twenty thousand years ago, near the peak of the last ice age, the world was a very different place than it is today. Temperatures were about seven degrees Fahrenheit cooler and the climate was, in most places, drier. In North America, all the ice-age creatures we know and love from *Ice Age*, the movie — mammoths, sloths, saber-toothed tigers—roamed through the plains and forests. In the West, you could walk from what's now San Francisco to the Farallon Islands. The Laurentide ice sheet, thousands of feet thick in places, covered most of Canada and the Upper Midwest and spread along the East Coast all the way to New York City. In Europe, it was dry land from London to Paris, and up north, from Scotland all the way over to Sweden. In Asia, you could walk from Thailand to Indonesia, and then take a boat down to Australia.

And people did. One wave of migration, as every American kid learns in middle school, brought humans across the land bridge between Asia and North America, thus laying the groundwork for the birth of Hollywood, Silicon Valley, and Ben & Jerry's ice cream. Exactly why and when human beings made the trip across the land bridge in the first place is much debated. Until recently, the best guess for the date of the arrival of humans in North America was 13,200 years ago. Many anthropologists believed it couldn't have been much earlier than that because although the land bridge was open, much of North America was covered in an ice sheet until then, making it virtually impossible for even the most intrepid early explorers to travel down into the interior of the continent.

But that narrative has been challenged. In 2012, Jessi Halligan, a young anthropologist at Florida State University, led a team of divers to explore the Aucilla River about seventy-five miles from Tallahassee. The Aucilla is a slow, dark, mysterious river that winds across northern Florida's limestone plateau toward the Gulf of Mexico. Earlier archaeologists had pulled up boatloads of bison bones, saber-toothed tiger bones, and mastodon bones and tusks, some with markings that looked like humans could have made them. During the ice age, the sea had met the land a hundred miles farther out, and the area where the Aucilla now flows was high savannah. Springwater bubbled and pooled in the limestone, creating watering holes where animals gathered to drink. As the seas rose and the water backed up, these watering holes filled with sediment, covering and preserving the bones of the animals that had died along the waterholes.

In May of 2013, Halligan's team made one of those scientific finds that change the way we view the world. In a sinkhole in the river, surrounded by mastodon dung, they found a two-faced knife that could only have been made by humans. More important, Halligan was able to precisely carbon-date the knife to 14,500 years ago.

14,500-year-old knife found in Florida's Aucilla River. *(Photo courtesy of the Center for the Study of the First Americans, Texas A&M University)*

That finding is important in a number of ways. First, it is indisputable proof that humans were hanging around in Florida a thousand years earlier than had been previously understood. There was other evidence that humans had been in North America earlier, including artifacts at archaeological sites in places as far-flung as Oregon and Chile, but none were as solid as this one. Second, it suggests that these early immigrants were more creative and resourceful than researchers had previously understood. "We know that until about twelve thousand, six hundred years ago, the route from Alaska down to the interior of the continent was blocked by ice," said Halligan. "The only way people could have gotten from Asia to this spot in Florida fourteen thousand, five hundred years ago that doesn't involve time travel or teleporting is if they came by boat." Halligan suggested they might have come down the West Coast, perhaps to Central America, then across the Gulf to Florida. If this is true, then these Paleolithic hunter-gatherers were building boats, understanding currents, navigating coastlines, and storing food and water. Of course, looking for evidence of this path down the coast is nearly impossible—many artifacts and campsites are now under three hundred feet or so of Pacific Ocean.

What is most important about this discovery—at least for the purposes of this book—is that the date of the double-sided knife corresponds with the sudden disintegration of the ice sheets at the end of the last ice age.

Scientists refer to the event as Meltwater Pulse 1A. It occurred just as the Earth was warming at the end of the last ice age. In coral reefs and other geologic sites around the world, scientists have seen that in the space of about 350 years, starting 14,500 years ago, the oceans began rising at a dramatic rate—more than a foot per decade. They know that this kind of sudden rise could only come from the collapse of a very big chunk of ice; the most likely candidates are the Laurentide ice sheet that covered North America and the glaciers of Antarctica. Scientists don't know the mechanism for the collapse, whether it was the sudden breaking of a giant ice dam in North America that was holding back meltwater from the Laurentide, or warm ocean water getting up under the ice sheets of West Antarctica. But the geological evidence for the event itself is indisputable. It happened.

Due to the flat topography of coastal Florida, the rising seas would have been particularly dramatic to anyone living there. Halligan estimates that the seas moved inland at a rate of five hundred to six hundred feet a year. That's a mile of coastline lost per decade—fast enough that you could almost watch the water come in while you gutted fish on the beach.

Halligan doubts that sea-level rise was the reason people abandoned the watering hole, since evidence so far suggests that the butchery at the site occurred over a very short period (they left no written accounts, which isn't surprising, since writing hadn't been invented yet). But whatever happened, it's clear that rising seas were radically reshaping the world they lived in. And they weren't the only ones who were dealing with it. At the time of Meltwater Pulse 1A, there were about three million people living on the Earth—nearly the population of Los Angeles today. They were living in small groups, making tools, hunting game, taking baby steps on the long ladder to modern life. What did they think about? What did they fear? Researchers can only make inferences from campsites, tools, and stray artifacts.

The most revealing clues, however, may come from the stories they left behind.

Nicholas Reid is an Australian linguist who studies the dying languages of Australian aborigines. Back in the 1970s, during his undergraduate days, he read a book called *A Grammar of Yidiɲ*, about a nearly extinct aboriginal language spoken in northern Australia. For years, one particular sentence in the book stuck with him: "It is, however, worth noting that a theme running through all the coastal Yidinji myths is that the coastline was once where the barrier reef now stands (as in fact it was some 10,000 years ago), but the sea then rose and the shore retreated to its current position." The idea lingered with Reid over the years. Was it possible that a ten-thousand-year-old event such as sea-level rise could be the basis for aboriginal myths?

In 2014, Reid mentioned the idea to a colleague, Patrick Nunn, a marine geologist who had studied sea-level rise in the Pacific. Nunn suggested that if the specifics in the myths were clear enough and detailed enough, they could be corroborated with geological data, allowing him and Reid to essentially date the origins of the myths.

Aboriginal societies have existed in Australia for around sixty-five thousand years, isolated until the European colonization of the continent in 1788. Australia was undoubtedly a hard environment to live in, and survival through the generations depended on passing down information about food, the landscape, and the climate. But that didn't necessarily mean the stories the Aboriginals told were, after thousands of years of retelling, in any sense "precise."

"It had long been assumed by linguists that the oldest oral stories are eight hundred years old—after that, any specific references in the stories are lost," Reid explained to me. "How could a story be told accurately, over and over again, for ten thousand years?"

Still, it was a fascinating possibility. Reid began reading aboriginal myths, many of them collected in the late nineteenth and early twentieth centuries by Western researchers. Without trying too hard, he found twenty-one examples of stories about sea-level rise. Each one was different, but they seemed to be chronicling a time when the sea was rising and the people who lived on the coasts and the islands were grappling with how to deal with it. In regions of Australia where the coastal land had a low topography, even a small rise in sea level would have claimed large chunks of land relatively quickly. "People must have been aware that every year the sea was getting higher," Reid said. "And they must have had stories from their fathers and grandfathers, and great-grandfathers, that the sea used to be out even further."

Here is one example:

In the beginning, as far back as we remember, our home islands were not islands at all as they are today. They were part of a peninsula that jutted out from the mainland and we roamed freely throughout the land without having to get in a boat like we do today. Then Garnguur, the seagull woman, took her raft and dragged it back and forth across the neck of the peninsula, letting the sea pour in and making our homes into islands.

This story is about the origin of the Wellesley Islands off the northern coast of Australia, and it parallels other stories from other parts of Australia. Along the south coast, stories written down early in colonial times told about when these areas were dry, a time when people hunted kangaroo and emu there, before the water rose and flooded them, never again to recede.

"There are numerous Aboriginal stories from this area about a time when the shoreline was further out 'where the barrier reef now stands,'" Reid told me. According to one of the stories, the barrier reef was the original coast at a time when a man called Gunya was living there. Gunya consumed a customarily forbidden fish, which made the gods angry. To punish him, they caused the sea to rise in order to drown him and his family. "He evaded this fate by fleeing to the hills but the sea never returned to its original limits," Reid said.

Another story collected from the Yidinji people of the Cairns area—now a coastal town that is a frequent jumping-off point for expeditions to the Great Barrier Reef—recalls a time when Fitzroy Island, which is now a mile or so off the coast, was part of the mainland. The story describes several named landmarks with remembered historical-cultural associations that are now underwater. According to Nunn and Reid, based on the details in the story, researchers can be almost certain that the people of this area occupied the coast "where the Great Barrier Reef now stands" during the last ice age, when it was a broad floodplain with undulating hills, bordered by steep cliffs—which are now islands like Fitzroy.

"Our expectation originally was that the sea level must have been creeping up very slowly and not been noticeable in an individual's lifetime," Reid told me. "But we've come to realize through conducting this research that Australia must in fact have been abuzz with news about this. There must have been constant inland movement, reestablishing relationships with the country, negotiating with inland neighbors about encroaching onto their territory. There would have been massive ramifications of this."

Still, the idea that these stories were a chronicling of actual events was remarkable. "If you are talking about ten thousand years, you are really talking about three hundred to four hundred generations," Reid said. "The idea that you can transmit anything over four hundred generations is extraordinary." But Reid believes a key feature of Australian Aboriginal storytelling culture—a "cross-generational cross-checking" process—could explain the stories' endurance through the millennia. In this process, a father will pass down the story to his sons—and the son's nephews and nieces will be responsible for ensuring that their uncle knows those stories correctly.

These stories, of course, tell us nothing about what these aboriginal tribes thought or felt about the seas rising around them. But they do capture how deeply significant and strange this experience must have been, how inexplicable.

*   *   *

The best-known flood story in the Western world, of course, is Noah's. In the Old Testament, the story is told of how Noah builds an ark and loads all the animals into it so they can survive the flood that God sends to cleanse the Earth. In God's view, there is just too much corruption and debauchery going on in his lovingly created paradise, and he has to do something about it. It's a powerful story of sin and redemption, but it's not original to the Old Testament. Most Biblical scholars believe that the story of Noah is based on an even earlier flood story in *The Epic of Gilgamesh,* which is the tale of the adventures of a Mesopotamian king that was written two thousand years before the Bible.

There is nothing in the Bible itself—or in *Gilgamesh*— that suggests these stories have anything to do with sea-level rise ten thousand years ago—or any other time. In both cases, the flood is caused by epic rains.

But two scholars think it's more complex than that. William Ryan and Walter Pitman, both geologists at Columbia University, argue that *Gilgamesh,* as well as the later Noah flood story, are based on a real event that occurred in the Black Sea about seven thousand years ago, when the seas were still rising at the end of the last ice age. At the time, the Black Sea was an isolated freshwater lake, cut off from the Mediterranean Sea by a high, narrow strip of land in what is now Turkey. Small groups of people lived on the fertile land around it, fishing from small boats and experimenting with growing crops for food.

As the ice melted, the Mediterranean rose higher and higher, and by about 5600 BCE, it had risen to a point where it was 500 feet above the Black Sea. Then the strip of land between them collapsed, and the seawater flowed over it. So much water poured in so fast it cut a flume—now the Bosporus Strait—280 feet wide and 450 feet deep. Ryan and Pitman calculated that ten cubic miles of water rushed through each day, two hundred times what goes over Niagara Falls, enough to cover Manhattan each day with water a half mile deep. The level of the lake rose six inches a day, inundating the deltas and invading the flat river valleys—moving upstream as much as a mile each day. "It's hard to imagine the terror of those farmers, forced from their fields by an event they could not understand, a force of such incredible violence that it was as if the collected fury of all the gods was being hurled at them," Pitman and Ryan have written. "They fled with family, the old and the young, carrying what they could, along with fragments of other languages, new ideas, and new technologies gathered from around the lake."

After two years, the lake water had risen 330 feet, until the lake was at the same level as the Mediterranean Sea. The people who had lived around the lake scattered to Europe and the Middle East, spreading their agricultural skills and know-how into the West and down into what became Mesopotamia, where stories of the flood became the basis for the flood story in *Gilgamesh*—and later, the Bible.

It's not a thesis all scientists accept. Liviu Giosan of Woods Hole Oceanographic Institution and colleagues from the University of Bucharest drilled cores in the area, examining the sedimentary data near where the Danube River empties into the Black Sea. They found evidence that Black Lake/Sea water levels rose only about half as much as Ryan and his colleagues proposed and would have drowned only about 800 square miles of land (about half of Rhode Island), rather than the 25,000 square miles (more than the entire state of West Virginia) that Ryan and Pitman suggested.

THE DELUGE.

"The Deluge," from 19th-century French artist Gustave Doré's illustrated edition of the Bible. *(Photo courtesy of Wikimedia Commons)*

However serious the Black Sea flood may have been, researchers will likely never know for sure whether or not it inspired the flood stories in *Gilgamesh* or the Bible. But it is certainly true that flooding was a frequent and destructive occurrence in the ancient world and a common metaphor for political and social dissolution. In both *Gilgamesh* and the Bible, the flood is a catastrophe—but it's also a cleansing, and a way of preparing the fallen world for a new order to emerge.

Unlike other ice-age mammals, humans had adapted to a changing climate and rising seas pretty well. One group who probably managed this better than most was the Calusa, a Native American tribe who lived in South Florida until they were wiped out by smallpox-bearing Europeans in the eighteenth century. To get better insight into how they lived, I visited Mound Key Archaeological State Park, which is on an island on Florida's Gulf Coast that was the ancient capital of the Calusa.

My guide was Theresa Schober, an archaeologist and former museum director who had been studying the Calusa for more than a decade. I met her at the boat launch at Lovers Key State Park near Fort Myers and we loaded our gear onto a sixteen-foot fishing boat that was piloted by a friend of hers. Schober, then forty-six, was tall and thin and muscular, with an infectious enthusiasm for Calusa lore. We sped across Estero Bay, dodging Jet Skis and other fishing boats beneath a sky that was full of the towering, tumbling clouds that Florida is so good at creating. From a distance, Mound Key looked like any other Florida island—low and green and peaceful. The only thing remarkable about it was that it was entirely artificial, an island built by the Calusa from their discarded seashells.

We approached the island through a tangle of mangroves, idling up a narrow channel that felt like a portal to another time. As Schober explained to me while we struggled out of the boat onto a small beach, *Calusa* meant "fierce people" in their language. No one knows exactly how long the Calusa lived in this region, but it was likely thousands of years. Their first European encounter was in 1513 with Ponce de León, the Spanish explorer who, legend has it, was in search of the Fountain of Youth. The Calusa attacked his ships and drove them away. Unwisely, he returned to the area nearly a decade later. The Calusa attacked again—this time, they shot de León with an arrow poisoned with the sap of the manchineel, an applelike tree that grows among the mangroves in Florida. The Spanish name for it is *árbol de la muerte,* or "tree of death," because the sap contains nasty toxins, including an organic compound called phorbol. De León died a few weeks later in Puerto Rico.

I dug my heel into the sand on the beach, thinking I might see an ancient oyster shell. "You'll have to go a little deeper," Schober joked. It was hard to believe that this entire island—all 125 acres—was created by generations of early Americans tossing oyster and mussel shells out the back door of their *palapas.* It was basically a well-engineered dump. Hunter-gatherers left these shell middens, as they are called, all over the world, from Australia to Denmark. In Florida, there are middens on both coasts and alongside most rivers. Many coastal middens are underwater or have been destroyed by development. But Mound Key, Schober explained, was in pretty good shape.

As we headed down a path through the mangroves, Schober told me that when the Spanish arrived, there were about a thousand people living on and around the island. But they were hardly isolated. They traded furs, food, glass beads, and other goods with neighboring tribes—there's even evidence that they canoed all the way to Cuba.

"Did they leave behind any art? Any stories?"

"Nothing," Schober said. "They were wiped out. These middens are it."

After hiking for twenty minutes or so, we came to what looked like a wide, shallow ditch cutting across the trail. "This was the grand canal," she explained. "The Calusa were very good at engineering with water. They built canals, with locks to control the water, as well as big water courts that functioned like a city square. Water was not something they resisted—it was deeply a part of their lives."

"The Pinelands Site," by Florida artist Dean Quigley. Early Floridians like the Calusa were well adapted to living with water. *(Illustration courtesy of Dean Quigley)*

Of course, the Calusa were not the only people who knew how to live with water. In New England, Native Americans lived in wood-framed houses that were covered with grass mats or bark so they could be taken down and transported by water. In parts of Newfoundland, a practice called launching is still common: when the waters rise or the shore changes, houses are dragged on wooden sleds to new locations. During the Revolutionary War, Tories escaping from Maine brought their houses with them to New Brunswick, where they still line the harbor. Houses on Cape Cod were also moved and recycled. One observer found that the residents thought of their "houses less as family seats, founded for the ages, than as temporary shelters, like the borrowed shells of hermit crabs, to be shifted about and exchanged, in location and function, as the need arose."

Much of this is forgotten knowledge now.

"I was living here in South Florida when Hurricane Charley came through," Schober said. Charley was a Category 4 hurricane that blew through the area in 2004. "People lost power, couldn't get gas for their cars. It was a total disaster for lots of people." Schober pointed out that the Calusa had to deal with hurricanes and storms too, but for them, it was probably not a big deal. "They just rebuilt their homes; it was part of their lives. If the storm changed the shoreline, that was how it was. If it blew over their huts, they could rebuild them. I don't think they had a sense of permanence, a sense that their world was settled. Their world was changing every day."

We left the canal and climbed a narrow trail through the mangroves to the top of the midden, which was about thirty feet high. In South Florida, that feels like Mount Everest. "The chief's house would have been up here," Schober explained. "The higher your house, the more prestigious it was—just like in cities today."

It was an oddly hopeful moment: here I was, standing on an artificial island built by Calusa a thousand years ago, one oyster shell at a time. Schober explained how the shells interlocked and calcified together over time, creating a solid structure that has survived for thousands of years. It was not just a monument to human ingenuity, but also a sign that living with water is something our ancestors have been doing for a very long time. Of course, the Calusa didn't have to worry about salt water corroding electrical wiring, or property values crashing, or nuclear power plants melting down as they got swamped by rising seas.

2. LIVING WITH NOAH

LIKE ANY RESPECTABLE Miami lawyer, Wayne Pathman owns an expensive sports car and lives in a big house on the water in the Sunset Islands neighborhood of Miami Beach. The Sunset Islands (there are four), like many of the islands that have arisen in Biscayne Bay, were created by digging up mud and piling it high and then surrounding it with a wall to keep the mud from washing away. Basically, it's the same thing the Calusa did for thousands of years. Many of the homes on the islands, which are just a few feet above sea level, go for $10 or $15 million and have stunning views of downtown Miami. Pathman lives just down the street from Philip Levine, the wealthy former mayor of Miami Beach, and not far from a $25-million Mediterranean Revival mansion that rocker Lenny Kravitz once owned. Pathman, who was in his early fifties at the time of this writing, grew up in Miami Beach and built his career handling land lease and zoning negotiations for Miami businesses and developers. In 2017, he became chairman of the Miami Beach Chamber of Commerce, where he has worked hard to get Miami's business leaders to understand the risks of sea-level rise and, among other things, stop building developments right on the water. "Noah was right," he told me at dinner one night. "When you talk about flooding, nobody listens. They all think it's not their problem. Have you taken a helicopter up and seen the cranes?"

I had not, but it sounded like a good idea to me. A few weeks later, I found myself up in the air with Sheryl Gold, another Miami Beach native and longtime community activist, and Roni Avissar, a former Israeli military pilot who is now a professor of atmospheric sciences at the University of Miami. We flew from a small heliport to the west of the city, near the Everglades, then swooped in toward Miami at low altitude. I could see boats cruising across Biscayne Bay, people sunbathing on rooftops. But Pathman was right: from the air, downtown Miami was a forest of cranes. Most of the construction was condo towers. A number of them were designed by rock-star architects like Norman Foster or Zaha Hadid and were architecturally interesting in an early-twenty-first-century postmodern sort of way. But from the helicopter, they all looked the same.

From the air, you could see how the city pushed up against the ocean. And it wasn't just new condo towers—it was hotels, hospitals, university buildings. They were all right on the shore, dangling their toes in the water.

Rachel Carson, author of *Silent Spring,* a book about the dangers of pesticides that inspired the modern environmental movement, tried to articulate the roots of our desire to live near water. She wrote about how life itself grew out of the sea, and how "each of us began his individual life in a miniature ocean within his mother's womb, and in the stages of his embryonic development repeats the steps by which his race evolved, from gill-breathing inhabitants of a water world to creatures able to live on land." Carson foresaw humanity making its way back to the sea, where, if it could not return to the ocean physically, it would "re-enter it mentally and imaginatively."

From the air, that was precisely what seemed to be happening in Miami.

A few billion years ago, Florida was part of Africa. When the Atlantic Ocean opened up, Florida was left behind, stuck onto the North American continent. It was just a big chunk of rock. Sea levels rose and fell over Earth's history, covering Florida with hundreds of feet of ocean water for millions of years, then exposing it again. Each time the water was high— and that was a majority of the time—the ocean was full of microscopic creatures that ate and shit and died. Their skeletons and excrement and shells drifted to the bottom, along with bits of coral and grains of sand and mud, most of which washed down from the Appalachian Mountains in the north and onto what is now Florida. Eventually, all this stuff went through a chemical transformation that cemented it together and turned it into limestone. That limestone grew thicker and thicker, an accumulation of excrement and skeletons and coral that is now three thousand feet deep in some places.

During the rise and fall of the seas, the water sometimes paused long enough to allow reefs to form, or to create little pearl-like grains called ooids. One particularly unusual set of conditions arose about 120,000 years ago, when sea levels were about twenty feet higher than they are today. Along the southern coast of Florida, shallow, warm, turbulent waters created what was essentially an ooid factory, where bits of shells and pellets of muddy shrimp excrement and coral could tumble around and acquire a fine coating of carbonate that gave them a pearly luster. When the ooids grew to about the size of large grains of sand, they settled to the bottom. (You can see the same process on certain beaches in the Bahamas today.) Over time, the ooids piled up, and, when sea levels fell again, cemented into ooid limestone. Eventually the ooids themselves weathered away, leaving behind an unusual limestone with holes in it. That pile of porous ooid limestone is now part of what is known as the Atlantic Coastal Ridge, which is roughly thirteen feet above sea level and runs from Palm Beach down to about Homestead. More than five million people live along that ridge today.

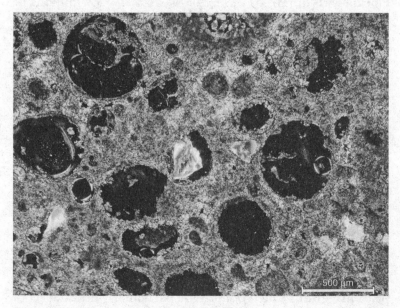

Close-up of Miami limestone with dissolved ooids. The dark areas are porous, allowing water to flow through. *(Photo courtesy of D. F. McNeill/University of Miami)*

In the pancake-flat topography of southern Florida, the emergence of the coastal ridge is a big deal. It prevented water from draining from the flat land to the west of the ridge, turning it into a swamp that became the Everglades. Eventually, a few rivers worked through low spots, allowing some water to flow out of the swamp (the Miami River, which cuts through downtown, is the biggest). Over time, the old reef eroded. Seeds lodged here and there. Hardy trees like pine and mahogany grew, and the ridge became a rocky highway between the swamp and the beach. The Tequesta, a Native American tribe on the east coast of Florida that was related to the Calusa on the west coast, used it for travel, as did panthers and deer and other dry-land-loving creatures of South Florida. And in 1890, it was here on the coastal ridge, right at the mouth of the Miami River, that a forty-one-year-old widow named Julia Tuttle bought a house that was once part of Fort Dallas, a remote military outpost built in the early nineteenth century. Tuttle fixed up the old house and made it a showplace—she was arguably the first person to grasp both the beauty of the landscape and the opportunities to get rich quick in South Florida.

Tuttle's home is long gone, of course. But there's a historic marker on the spot. If there is a dead center of today's Miami, the middle of the crane forest, this is it. On either side, towering condos look out over Biscayne Bay, toward the port and Miami Beach. As I walked along the shore one day, enormous yachts motored up the river, some blaring Drake or Kanye West, festooned with men and women in skimpy bathing suits. I flashed back to pictures I'd seen of Tuttle, "the Mother of Miami," who first persuaded Henry Flagler, the exceedingly rich cofounder of Standard Oil, to build a railroad down to this godforsaken place. The story of her sending orange blossoms in 1896 to Flagler, who had a house in Palm Beach, and persuading him to extend his railroad from Palm Beach to Miami is the founding myth of the city.

And once Flagler's railroad arrived—well, you know what happened. The twentieth century happened.

Although they didn't know they were living atop an old coral reef, the locals called the high rocky ridges reefs and the lower sandy, soggy areas glades or sinks. Homesteaders liked the glade because it did not require clearing and the soil was good for raising vegetables—especially winter vegetables that grew to maturity before the summer rains. Once you cleared out the pine and mahogany, the rocky ridges turned out to be good for citrus. The elevation of the ridges also gave protection from the biggest threat settlers faced there: water. In South Florida, that boundary between water and land was always amorphous.

Exhibit A: the memoirs of George Merrick, who was the founder of Coral Gables, a master-planned community south of Miami, where much of the Florida gentry (including former governor Jeb Bush) now lives. In 1901, Merrick was a fifteen-year-old boy living on the family homestead near Miami. The house was up on the coastal ridge, but that didn't stop the water when torrential rains hit that year. "We were living with Noah," Merrick later wrote in an account of his life on the homestead. All the land below the coastal ridge flooded. The Merricks' vegetable garden disappeared under six feet of water. Roads became impassable, and on the low ones, the water was so deep that some wagons floated away.

Inside their "ark," as Merrick called their house, it was even worse. As the water rose, the family pulled boards off the barn and nailed them to the cabin floor to elevate it above the floodwaters. Armies of roaches and other insects sought refuge inside the house. "The more we killed, the more the roaches seemed to reappear, as if raining from the skies," Merrick wrote. The frogs invaded too: "There was something horrifying in the clamor that unceasingly enveloped the cabin. The din was as if every rain drop, as ceaselessly they fell, gave birth to a new voice; a new croak, gurgle, gurk, grackle and shriek." Alligators swam in from the Everglades, bellowing, devouring half-drowned rabbits. Merrick and his father floated wood in from the barn for the cookstove on a makeshift raft. Then the manure pit overflowed, causing Merrick's father's feet to break out in wormlike red eruptions, which Merrick's mother tried to cure by pouring iodine into the open wounds.

Early settlers like the Merricks understood that if civilization was going to continue, something had to be done about the water. More important, these pioneers figured out that draining the Everglades was the equivalent of creating free land. By 1909, dredging of the Miami Canal had begun, and what followed was perhaps the most rapid, most dramatic, most heedless remaking of a landscape that humans had ever attempted. When this massive water-diversion scheme was finished, thousands of acres in the Everglades had been drained dry and opened up to the speculators.

And speculate they did. The taming of the Everglades unleashed a land boom that was unlike anything seen before in America. "From the time the Hebrews went into Egypt, or since the hegira of Mohammed the prophet, what can compare to this?" one newspaper mused. The pilgrims included celebrities like the boxer Gene Tunney, the actor Errol Flynn, and businessmen such as Alfred du Pont, J. C. Penney, and Henry Ford. But ordinary Americans also headed to Florida to get rich, enjoy vacations, or retire in a sunny climate. "The suntan, once a symbol of labor, became a symbol of leisure," Michael Grunwald writes in *The Swamp: The Everglades, Florida, and the Politics of Paradise.*

But even then, there were dissenters. As Grunwald points out, Charles Torrey Simpson, the most eloquent of Florida's preservationists, suggested a new ethic in which Floridians no longer considered themselves superior to nature and stopped trying to tame it and exploit it:

> There is something very distressing in the gradual destruction of the wilds, the destruction of the forests, the draining of the swamps, the transforming of the prairies with their wonderful wealth of bloom and beauty—and in its place the coming of civilized man with all his unsightly constructions, his struggles for power, his vulgarity and pretensions. Soon this vast, lonely, beautiful waste will be reclaimed and tamed; soon it will be furrowed by canals and highways and spanned by steel rails. A busy, toiling people will occupy the places that sheltered a wealth of wildlife.... In place of the cries of wild birds will be heard the whistle of the locomotive and the honk of the automobile. We constantly boast of our marvelous national growth. We shall proudly point someday to the Everglade country and say: Only a few years ago this was a worthless swamp; today it is an empire. But I wonder quite seriously if the world is any better off because we have destroyed the wilds and filled the land with countless human beings.

Geologically speaking, Miami Beach is a new kid in town. Three thousand years ago, around the time the Great Pyramids were being built, the sandbar that we now call Miami Beach began to form on the platform of ooids off the coast. The sand (most of which was ground-up rocks from the fast-eroding Appalachian Mountains) began to collect in the shallow water. Mangrove seeds washed up. Insects arrived. By the late 1800s, it was a dense tangle of mangrove and palmetto, rattlesnakes, rats, mosquitoes, and other insects. "Virgin jungle crept right down to the beach, a jungle as dense, forbidding and impenetrable as only the tropics can produce," one early visitor wrote. "It was impossible to proceed more than a few feet into its depths without hacking a path with a machete."

For most of human history, few people thought of beaches as welcoming places. In fact, for early European explorers, the beach had been a place to be avoided. It was a place where you might land your boat, but otherwise it was a treacherous zone that was associated with death and disease and that marked the line between civilization and nature. When the English, Dutch, and French began to settle the New World in the seventeenth century, they weren't scouting potential beach resorts. They wanted timber, fur, and fish.

In Europe, the first people to occupy the beach and build houses facing the sea were upper- and middle-class landlubbers who were convinced of the therapeutic quality of sea air and water. In the mid-eighteenth century, lured by extravagant promises of miracle cures, English invalids and hypochondriacs began gathering in places like Brighton, on the English Channel. They frolicked in newly invented bathing machines and spent hours beachcombing and sightseeing, activities that were completely alien to the locals, who, as novelist Jane Austen observed, avoided the water except when making a living from it: "Sea views are only for urban folk, who never experience its menace. The true sailor prefers to be landlocked rather than face the ocean."

By the mid-nineteenth century, people began building houses and hotels closer to the shore. Newly built piers in places like Atlantic City, New Jersey, were essentially extensions of land and carried visitors over rather than onto the beach itself, giving them a safe vantage point from which to look both to sea and back to land. For these early vacationers, the pull of the beach was its very emptiness and "cleanliness"—no shipwrecks, no dead bodies washing up, no sign of dirty industrial life. One French sociologist called this the "aesthetic conquest of the shore by the vacation ideology."

By the 1870s, Atlantic City was a full-fledged beach resort, and Coney Island had roller coasters and luxury hotels. But Miami Beach was still just a tangle of mangroves and mosquitoes. In 1876, the US government built what may have been the first permanent structure on Miami Beach — the Biscayne House of Refuge. Intended for victims of shipwrecks, of which there were many on this treacherous coastline, it was stocked with provisions, clothing, blankets, and first aid equipment. There must have been some romance to the place, however, because records show that the wife of Jack Peacock, keeper of the house, bore children there in 1885 and 1886.

In the 1890s, around the same time Tuttle was coaxing Flagler to build a railroad to Miami, the first speculators arrived on Miami Beach. The most important one was a New Jersey farmer named John Collins (the namesake of today's Collins Avenue, a main thoroughfare in Miami Beach), who bought several hundred acres of the island for next to nothing, cleared some land, and planted 38,000 coconut trees, thinking he would make a fortune. It turned out to be a failure. He then planted 2,945 avocado trees, and they did a little better. But the most notable thing Collins did was start building a bridge from Miami, across Biscayne Bay, to the sandbar, thinking it would draw in people and make land more valuable. And the most notable thing *that* did was capture the attention of Carl Fisher, a daring speed-obsessed Midwestern entrepreneur who got rich from his patent and manufacturing of the first mass-produced automobile headlight and helped create the Indianapolis Motor Speedway and the first transcontinental highway. But mostly, Fisher was a huckster and a promoter who saw the potential of transforming Miami Beach into America's winter playground. Fisher gave Collins the money to finish the bridge in exchange for several hundred acres of seemingly worthless swampland.

Carl Fisher in Miami Beach, 1923. *(Photo courtesy of Miami Beach Digital Archives)*

It was an insane project, as Fisher's wife, Jane, immediately understood when he took her out to inspect his new real estate.

"Creatures that made me shudder were lying in wait on the branches of the overhanging trees," she recalled. "The jungle was as hot and steamy as a conservatory. The mosquitoes were biting every exposed inch of me. I refused to find any charm in this deserted strip of land. But Carl was like a man seeing visions. He picked up a stick, and when we reached the clean sand he began to draw a plan. I know now that he was seeing Miami Beach, in its entirety, rising from that swamp."

Fisher hired hundreds of black laborers to hack away the palmetto and mangroves. Then he filled in the marshy land with sand dredged from the bottom of Biscayne Bay. He brought in colossal steel-hulled dredgers, one of which had on board its own complete machine and repair shop and its own ice plant. It had a 1,000-horsepower engine that pushed the sandy bay bottom to the surface through a twenty-inch pipe. It could move 20,000 cubic yards of fill in twenty-four hours.

The fill was a mixture of sand, mud, and marl, resembling "sloppy Cream of Wheat," as one observer described it. Men worked in mud and slime up to their knees, wearing long boots to guard against snakebites. One of the first jobs was to transform Bull's Island, where the Collins Bridge ended, so that the road could be built out to the beach. One of the dredgers pumped 300,000 cubic yards of sand on Bull's Island in four days' time, creating acres of land. For aesthetic reasons, the island was renamed Belle Isle in 1914 (today, Belle Isle is populated by condo towers, as well as the Standard, a hip hotel and spa where you can buy vibrators in the hotel gift shop). The "soup" had to drain and thicken before anything could be done with it. For six months, it was left undisturbed so that the algae and marine life could dry or rot away and the smell could dissipate.

Hard borders had to be built to keep it all from washing

131-13. 1923—Filling the polo fields; the Nautilus Hotel under construction in the distance.

Hi - 102

Dredger pumping "soup" up from the Biscayne Bay to make more room for development on Miami Beach. *(Photo courtesy of Miami Beach Digital Archives)*

away. Pile drivers sank supports for pilings, to which wooden timbers were attached with steel cables. Thousands of tons of rock were loaded onto barges in the center of the state, then floated down a drainage canal that went from Lake Okeechobee to Miami and across Biscayne Bay, where they were unloaded at a wharf and hauled by mules around Miami Beach. As one observer put it: "The bayside gradually began to look as if it were covered by a glittering, hard-packed layer of snow standing five feet above the tide." To keep the new land from blowing away, mulch from the Everglades was brought in by barge, spread, and graded. Jane Fisher remembered: "Hundreds of Negroes, most of them women and children, crawled on their hands and knees over the earth pushing ahead of them baskets of grass from Bermuda." Gradually, the wild island was tamed, made suitable for people motoring down from the Midwest.

A few miles up the coast, in Fort Lauderdale, Charles Green Rhodes, a West Virginia coal miner from a family of twelve, dredged a series of canals along the New River, then used the fill to create rows of small peninsulas that stretched into the river like grasping fingers. This way every lot on every street backed up to a canal and thus could be sold as waterfront. Rhodes named the development the Venice of America and made millions. His canal-building technique, dubbed *finger-islanding*, quickly caught on, and it took years before anyone realized that the canals became stagnant cesspools and the development destroyed the habitat of manatees and other wildlife. Or that the houses would sink into the muck. Or that, in the twenty-first century, as sea levels started to rise, the phrase *the Venice of America* would take on an entirely different meaning.

There was so much money to be made it was almost comical. A veteran who had swapped an overcoat for ten acres of beachfront after World War I found the property worth $25,000 during the boom. A screaming mob snapped up 400 acres of mangrove shoreline in three hours for $33 million. "Hardly anybody talks of anything but real estate, and... nobody in Florida thinks of anything else in these days when the peninsula is jammed with visitors from end to end and side to side — unless it is a matter of finding a place to sleep," said the *New York Times*. In 1916, Fisher's company sold $40,000 worth of real estate; nine years later, in 1925, the company sold nearly $24 million. By then, there were 56 hotels with 4,000 rooms in Miami Beach, 178 apartment houses, 858 private residences, 308 shops and offices, 8 casinos and bathing pavilions, 4 polo fields, 3 golf courses, 3 movie theaters, an elementary school, a high school, a private school, 2 churches, and 2 radio stations.

In 1925, humorist Will Rogers described Fisher as "the first man smart enough to discover that there was sand under the water. So he put in a dredge, and he brought the sand up and let the water go to the bottom instead of the top. Carl discovered that sand would hold up a real estate sign, and that was all he wanted it for."

Of course, the boom went bust. The Internal Revenue Service investigated Florida speculators, the Better Business Bureau exposed Florida con artists, journalists touted Florida scandals. And then, as if Mother Nature were getting her revenge for the blasting and draining of South Florida, at about 2 a.m. on September 18, 1926, a Category 4 hurricane slammed into Fisher's newly minted paradise. "Miami Beach was isolated in a sea of raving white water," journalist Marjory Stoneman Douglas wrote. Winds hit 128 miles per hour, turning utility poles into flying spears. Roofs Frisbeed off buildings. A ten-foot storm surge flooded Miami Beach. Homes floated off their foundations. When the water retreated, the streets were covered in sand, as were lobbies of swanky oceanfront hotels. The hurricane's final toll: 113 people dead, 5,000 homes destroyed or damaged.

In retrospect, none of this was surprising. The whole DNA of the Florida boom had been about making a quick buck. Nobody was thinking about resiliency; nobody was pausing to consider the consequences of building a city on the water right in the middle of a well-known hurricane path. The houses were flimsy, the wiring bad, the bridges fragile, the roads right along the water. Who cared? This was Miami Beach. There was no government oversight, no long-range planning. The only future that mattered was the next cocktail, the next beautiful sunset, the next riff in the jazz club.

And it wasn't just in Miami where flimsy construction proved to be dangerous. The 1926 hurricane pushed water through a poorly built earthen dike that held in Lake Okeechobee in the central part of the state, flooding the farmlands below the lake and killing four hundred people. Two years later, another hurricane hit, taking out an even bigger part of the still-flimsy dike and sending a fifteen-foot tsunami through the same farmland. This time, more than twenty-five hundred people died, mostly poor blacks who drowned in the vegetable fields of the Everglades.

Eventually, changes were made. The City of Miami passed the first building code in the United States (it later became the basis for the first nationwide building code). It required roofs to be bolted down, building frames to be securely attached to foundations, windows to have hurricane glass. The building codes have been upgraded over time, and Miami—like most modern cities—is far more resistant today to hurricanes than ever before.

But the legacy of cheap building in South Florida persists. You could see it in the devastation caused by Hurricane Andrew in 1992, which flattened Homestead, a region just south of Miami: $26 billion in damage, 65 people dead, 250,000 homeless. During my travels in South Florida, I learned that the concrete used in the much-beloved Art Deco buildings in South Beach was often mixed with salt water or salty beach sand, which can cause the rebar within the concrete to corrode over time and greatly weaken the

concrete. While renovating a South Beach hotel, one architect I know discovered that the structural walls were so weak
that you could practically knock them down with a hammer.
Instead of continuing with the renovation, they had to tear
down everything but the façade and rebuild it to modern
standards. How many other old Miami buildings suffer from
the same weakness? Leonard Glazer, an electrical engineer
and lighting designer who worked in Miami Beach in the
1950s and 1960s and was responsible for the spectacular
lighting at the Fontainebleau and the Eden Roc hotels,
laughed when I asked him about building codes: "There was
no code! Or if there was, nobody paid any attention to it.
Your job was to do the work quick, and do the work cheap. In
Miami Beach, nobody was thinking about the long term."

The 1926 hurricane also laid bare the combination of
boosterism and denial that drove — and still drives — South
Florida politicians and business leaders. Carl Fisher first artic-
ulated the Florida dream of sunshine, beaches, and fun (or,
as one Florida journalist wrote in 2016, "Miami Beach has
based its economy on the fact that strangers love to visit and
get drunk there"). Fisher used elephants and girls in skimpy
bathing suits and teams of PR pros to manufacture paradise.
Real estate was always a good investment; the sex was always
better, the climate always perfect. And hurricanes? In 1921 a
newspaper ad inviting tourists and investors to Miami Beach
promised there was "practically no danger from summer
storms." Of course, summer storms and hurricanes had been
sweeping through South Florida for thousands of years.

After the hurricane hit, denial went into overdrive. Miami
business leaders were terrified that hurricane publicity would
scare away visitors and investors. Florida's leaders minimized

the damage, denying reports of devastation as rumors and exaggerations, openly discouraging relief efforts. When a city editor at the *Miami Herald* filed a story reporting $100 million in damage, his boss ordered the losses reduced to $10 million. Miami's mayor declined offers of outside aid, and the governor insisted that life was quickly returning to normal. The chairman of the Red Cross charged that "the poor people who suffered are regarded as of less consequence than the hotel and tourist business in Florida," but official spin continued to portray the storm as a minor inconvenience in paradise. One booster took out full-page ads in the *Herald* pointing out that Florida was still perfectly positioned for growth, that the big blow was nothing compared to floods in the Midwest, "winter diseases" in New England, or earthquakes in California: "Sure, some lives were lost in the hurricane, but hurricanes come only once in a lifetime."

Carl Fisher and other early Florida developers may have been greedy quick-buck artists, but they knew something about defeating nature in pursuit of pleasure. They understood that no matter how mosquito-and-alligator-infested a swamp may be, you can drain it, dredge it, and make a fast dollar. No matter how big the hurricane, you can always rebuild. More than any place in America, South Florida has been an expression of the technological dominance of twentieth- and twenty-first-century life: it is a world created by dredgers, cooled by air conditioning, powered by nuclear energy, dominated by cars, sanitized by insecticides, glamorized by TV and the Internet. It is a place that has been habitable only if you believe the premise that nature—the heat, the bugs, the alligators, and most of all, the water—can be tamed.

3. NEW CLIMATE LAND

THE JAKOBSHAVN GLACIER on the west coast of Greenland is the fastest-moving glacier on Earth. It's also one of the most photographed, drawing climate paparazzi from around the world. If you've seen images of enormous chunks of ice falling off the sapphire-blue face of a glacier in news reports or documentaries, you've probably seen the Jakobshavn. It's the Kim Kardashian of glaciers—fast and unpredictable but symbolic of changes in our world in a way that makes it difficult to ignore. On my flight from Copenhagen to Kangerlussuaq, a town in southern Greenland that is the main launching point for visits to the ice sheets, I was, I admit, excited about the prospect of seeing the famous glacier up close.

My guide for the trip was Jason Box, an American climatologist who was working for the Geological Survey of Denmark and Greenland, a quasigovernmental agency that has a keen interest in understanding what is going on with the

Greenland ice (Greenland, a former province of Denmark, is still a territory of the Danish kingdom). Box is a young maverick scientist and Greenland ice junkie who got a lot of attention in 2012 when he publicly predicted just weeks before the summer melt season began that Greenland would experience a record-breaking year for ice melt. And he was right. During the summer of 2012, a heat wave warmed the Arctic much more than climate models had predicted, and the Greenland ice started melting like ice cream on a summer sidewalk. Usually, only the lower elevations of Greenland melt during the summer; in 2012, melting was recorded on the entire surface of the ice sheet, including the highest elevations. Aerial photos showed rivers of luminous blue water flowing across the surface of the ice sheet, disappearing into what scientists call moulins, which are basically big holes in the ice where the rivers waterfall down to the interior of the glacier. The melt of 2012 attracted international media attention, and a YouTube video of a tractor being washed down a river swollen with meltwater in Kangerlussuaq attracted millions of viewers.

When I met Box at the Kangerlussuaq airport, he had that scientist-in-the-wild look—scruffy, windblown hair, a big black duffel bag over his shoulder. With his goatee and gently rebellious manner, Box has the air of a skate punk made good. After chatting for a few minutes, he and I headed out into Kangerlussuaq. The town, such as it is, sprang up around the runway, which was built by the US military during World War II—and it still feels like a military base, with Quonset huts transformed into shops and hotels. I was surprised by the number of middle-aged people in town wearing new North Face jackets. "Disaster tourism," Box explained. "They're here to say goodbye to the ice sheets."

We walked down the main street, where I got my first look at Greenland—the high, rolling hills reminded me, oddly, of the hills north of San Francisco. Of course, in Greenland, there are no trees—and at least in June, when I was there, there was nothing else green either. Even though it was early summer, it was still chilly enough that we needed to wear light down jackets. There was no ice in view, but Box said there were plenty of glaciers just over the horizon. "You can hike to them in a few hours," he explained. Surprisingly, I could smell the ice in the air, as if someone had left a huge freezer door open.

We checked into what Box called the "science hotel" on the edge of town. It was basically a Quonset hut outfitted with beds and cable TV. We went for a quick evening walk, and Box showed me the Watson River where the YouTube of the tumbling tractor had been filmed during the 2012 melt. The river was about thirty feet wide now, docile as a backyard brook, gray with glacial silt. During the peak of the 2012 heat wave in Greenland, the Watson had been a raging torrent, flowing with ten times more water than the River Thames in England.

As we walked around town, Box explained that his goal on this trip was to test some unorthodox ideas about why Greenland was suddenly melting. Among other things, he believed it might be related to the darkening of snow caused by soot drifting down onto the ice sheet from coal-fired power plants in China and wildfires in the American West and northern Canada, as well as by dark-colored algae and bacteria growing on the surface of the ice. "Dark snow absorbs more heat than clean snow, which makes it melt faster," Box explained. "How big is the impact in a place like Greenland? I don't know. But it could be significant. And it's one of many things that are not in the climate models right now."

*   *   *

Most of the water that will drown Miami and New York and Venice and other coastal cities will come from two places: Antarctica and Greenland. Often you hear about the disappearance of the snows on Mount Kilimanjaro or the glaciers in Patagonia, but in the context of drowning cities, land-based glaciers won't contribute much. What really matters is what happens on the two big blocks of ice at either end of the Earth.

The risks in Greenland and Antarctica, as scientists understand them, are very different. Antarctica is about seven times bigger than Greenland and contains much more ice. If the whole continent were to melt (a scenario that would likely take thousands of years), it would raise the Earth's sea levels by about two hundred feet. If all of Greenland were to go (a scenario that could take significantly less time), it would raise sea levels about twenty-two feet. To put the volume of water we're talking about here into perspective, if all seven billion human beings on the planet suddenly jumped into the ocean, it would raise sea levels about one hundredth of an inch. Right now, melting from Greenland contributes roughly twice as much to current sea-level rise as Antarctica—but that may change in coming decades.

The Arctic, where Greenland is located, is one of the fastest-warming places on the planet. Not surprisingly, the main issue here is melting on the surface of the ice sheets, which is driven not only by warmer air temperatures but also by the amount of moisture in the air, the speed and direction of the winds, the cloudiness of the skies, and, as Box hypothesized, how much the surface of the ice has been darkened by soot or organisms like bacteria and algae.

Antarctica, in contrast, is the coldest place on Earth. East Antarctica, where the biggest ice sheets are located, is particularly frigid. Surface melting is not an issue here. But that's not the only way an ice sheet can disappear. In Antarctica, scientists are more concerned about warming ocean water melting glaciers from below and destabilizing the entire ice sheet. This is particularly worrisome in West Antarctica. Many of the biggest glaciers there, including Thwaites, which is roughly the size of Pennsylvania, are what scientists call marine-terminating glaciers, because large portions of them lie below sea level. Subtle changes in currents around West Antarctica have brought warmer water to the region. The change is small, but it's enough to increase the melt rate on the underside of these glaciers. Floating ice shelves, which grow like fingernails where the glaciers meet the ocean, are particularly vulnerable to melting from below. If the waters continue to warm, they are likely to fracture and break off. The crack-up of these floating ice shelves will not in itself raise sea levels (just as ice melting in a glass doesn't raise the level of liquid). But they play an important role in buttressing, or restraining, the glaciers behind them. If they collapse, the glaciers—some of them are ten-thousand-foot-thick mountains of ice—will be free to slide into the ocean.

Another factor that increases the risk in West Antarctica is the shape of the continent itself. If you had X-ray vision and could see though the ice, you would see that the ground below the ice sheet in West Antarctica is a reverse slope that has been depressed by the weight of the glaciers over millions of years. "Think of it as a giant soup bowl filled with ice," Sridhar Anandakrishnan, a polar glaciologist at Penn State University, told me. In this analogy, the edge of these glaciers sits perched on the lip of the bowl, and that lip is a thousand feet or more below sea level. Behind the lip, the terrain falls away on a downward slope for hundreds of miles, all the way to the Transantarctic Mountains, which divide East and West Antarctica. At the deepest part of the basin, the ice is more than two miles thick.

Some scientists fear that if the ocean around Antarctica continues to warm and the ice shelves collapse, these big glaciers may slip off their grounding lines and begin retreating backward down the slope, like "a ball rolling downhill," Ohio State glaciologist Ian Howat explained. The deeper down the slope the glaciers retreat, the taller and more unstable the cliffs at the calving front become, and the faster they can fracture and fall into the sea, eventually leading to what scientists call a runaway collapse of the ice sheet. And that, of course, would raise sea levels—fast.

In early 2017, a 100-mile-long crack opened in the West Antarctica ice sheet. *(Photo courtesy of NASA)*

The melting of Greenland and Antarctica will not be felt equally around the world. Paradoxically, the melting of Greenland will have a bigger impact in the southern latitudes;

the collapse of glaciers in Antarctica will be felt more in the northern latitudes. Scientists call this regional effect finger-printing, and it's a consequence of the way gravity spreads the water around the Earth as it spins. In both Greenland and Antarctica, the ice sheets melt and their mass gets smaller, which reduces their gravitational pull on the water around them. This causes the sea levels in the immediate area to fall—but that falling water pushes the water higher on the opposite side of the Earth. So when Greenland melts, it has a disproportionately bigger impact on Jakarta than on New York; when Antarctica melts, it has just the opposite impact. For example, the collapse of the big glaciers in West Antarctica would cause an average global sea-level rise of about ten feet. But in New York City, thanks to the pull of gravity, the water would rise thirteen feet.

"It's hard to get your mind around how fast the Arctic is changing," Jennifer Francis, an atmospheric scientist at Rutgers University, told me before I left for Greenland. According to NASA, Greenland is losing three times as much ice each year as it did in the 1990s. Between 2012 and 2016 alone, a trillion tons of ice vanished—enough to make a giant ice cube that is six miles on each side (that's taller than Mount Everest). Not so long ago, the Northwest Passage, the storied northern route from the Atlantic to the Pacific Ocean, required an icebreaker ship to navigate it. During the summer of 2016, seventeen hundred people cruised through the passage aboard the *Crystal Serenity,* a diesel-powered luxury ship complete with multiple swimming pools, movie theaters, and a crew of six hundred. By 2040, the summer sea ice in the Arctic is likely to vanish entirely—you'll be able to windsurf at the North Pole.

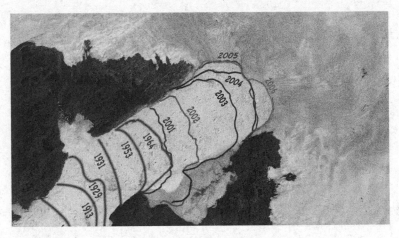

In the last 100 years, the Jakobshavn Glacier has made a dramatic retreat. *(Photo courtesy of NASA)*

In the past twenty years, the Arctic has warmed by more than three degrees Fahrenheit, roughly twice as fast as the global average. As the ice melts, the region's albedo, or reflectivity, changes. Clean, fresh snow is one of the most reflective substances known in nature, reflecting away more than 90 percent of the sunlight that hits it. But as the ice softens, its structure alters, lowering the reflectivity and absorbing more heat. As it melts away, more water and more land are exposed, both of which are darker, and both of which absorb still more heat. This in turns melts more ice, creating a feedback loop that can accelerate quickly.

This dramatic change in the Arctic may be causing ripple effects throughout the Earth's climate system. For example, some research has suggested a connection between the Arctic sea ice decline and the intensity of California's recent record drought (although the connection is not definitive). Other research has suggested that warming in the Arctic is reducing the temperature contrast between the Arctic and the tropics, causing wind patterns in the Northern Hemisphere to slow down. The result has been more summer climate extremes, including the deadly 2003 European heat wave and severe flooding in Pakistan in 2010.

Of course, scientists have understood the basic physics of reflectivity for a long time. But the behavior of ice sheets, which is notoriously hard to capture with conventional climate models, can come very close to chaos theory—where small changes in, say, the path of the jet stream or the amount of cloud cover can lead to enormous effects. The Great Melt of 2012 was an example of this. "Scientists didn't expect to see a total melt of Greenland for decades," said Michael Mann, the director of the Earth System Science Center at Penn State. "When it happened, you had to wonder—what is missing from our models? Is there some basic physics that we don't understand—or is there a human factor that we are not calculating, like the effects of soot on the snow?"

As every schoolkid knows, a water molecule is made up of two parts hydrogen and one part oxygen. Hydrogen was formed during the Big Bang roughly 14 billion years ago, while oxygen—a more complex element—was forged sometime later in the superhot interiors of stars. When stars died and went supernova, the explosion blew their elements into space, where oxygen and hydrogen mingled to form water.

The universe is awash in water. Scientists recently found a giant cloud of water surrounding a black hole 12 billion light-years from Earth. In our solar system alone, the interiors of Jupiter, Saturn, Neptune, and Uranus have enormous quantities of the stuff. Mars has ice caps on its poles, just like Earth, as well as belts of glaciers in its southern and northern latitudes (in fact, scientists have calculated there's enough water in glaciers on Mars to cover the entire surface of the planet with three feet of ice). The moons of Saturn and Jupiter have oceans beneath their icy surfaces. What makes Earth's water unusual is that it exists in a glorious liquid state between ice and vapor, and it's not too salty or too acidic or too alkaline. And we have a lot of it. The Earth's oceans cover about 70 percent of the surface of our planet. Without this vast cache of water, not only would there be no sushi and no kayaking and no warm showers after a hard day's work, but life as we know it would not exist. Life was born in the water and evolved there for billions of years before the first fish crawled up on the beach and set up camp on dry land.

What's not very well understood is where all the water on Earth came from. The most popular explanation is that it accumulated from icy comets and asteroids that battered the planet during the first billion years or so of its existence. That's a lot of dirty snowballs flying in from outer space, but it's possible. Another theory is that at least some of the water hitched a ride on the grains of dust that glommed together to form the Earth 4.6 billion years ago. Wherever the source was, scientists know the amount of water on Earth has been fixed for billions of years. It just gets rearranged, depending on the temperature of the planet.

The coming and going of ice ages, an accepted part of Earth's history for everyone except the most literal-minded creationists, wasn't well understood until the 1940s, when a

Serbian engineer named Milutin Milankovitch hypothesized that wobbles in the Earth's orbit altered the amount of sunlight that hit the planet at regular intervals, causing just enough variation in the Earth's temperature to trigger ice ages. As ice grew every 100,000 years or so, locking up more water, the seas fell. As the ice melted, the seas rose. If you watch a speeded-up visual rendering of the Earth over millions of years, the ice comes and goes in rhythm. It looks like the planet is alive and breathing.

The day after I arrived in Kangerlussuaq, we were supposed to fly up to Ilulissat, a small town on the coast that has become a scientific Mecca because it sits at the foot of the ice sheets. But the helicopter pilot Box had hired to get us out onto the ice — helicopters are like bicycles for Greenland ice scientists — was nowhere to be found. And without a chopper, it's virtually impossible to get out onto the glaciers to begin fieldwork.

So Box and some of his colleagues spent the day knocking around Kangerlussuaq. Box was clearly impatient with the delay — the clock was ticking, and every hour we spent on solid ground felt like an hour wasted. But we had to eat, and that evening, Box and I walked a mile or so to a small restaurant on the edge of a meltwater lake that served traditional Greenlandic fare like musk ox steak, smoked halibut, and whale carpaccio.

Over dinner, Box and I talked about the Big Melt of 2012. The fact that the real-world melt happened so much faster than models had predicted meant that something was missing from the models. But what? Did the wavering jet stream bring a heat wave to the region? Was it the heat-trapping properties of low clouds? Perhaps. But Box, who was working on what he called a "unified theory of glaciology," believed that soot-and-bacteria-darkened snow was a powerful overlooked factor. "It's going to take years to put all this together," he explained. "Unfortunately, given the rate at which the world is changing, those are years we don't really have."

Part of Box's charm, I discovered, is that he is not afraid to paint in broad strokes and is very conscious that his real audience is not other scientists but the general public, who, as he sees it, have been betrayed by the hesitance of scientists to make bold predictions. For Box, this is not a problem. For instance, in 2009, he announced that the Petermann Glacier, one of the largest in Greenland, would break up that summer— a potent sign of how fast the Arctic was warming. Box even led a scientific expedition to place instruments on the remote glacier so he could better track its disintegration. Most glaciologists thought he was nuts—especially after the summer passed and nothing happened. In 2010, however, Petermann began to calve; two years later, it was shedding icebergs twice the size of Manhattan.

"I like ice because it's nature's thermometer," Box told me over musk ox pizza. "It's not political. As the world heats up, ice melts. It's simple. It's the kind of science that everyone can understand."

At first, real-time monitoring of ocean levels had nothing to do with sea-level rise. It began with simple tide gauges in the early nineteenth century. In 1807, Thomas Jefferson requested that the US government begin a systematic survey of the coastline to map the new nation and facilitate maritime commerce. Because the coastline changes with the tides, that meant the surveyors also needed to begin measuring changes in the coastal water levels. (The oldest continuously recording tide gauge in the world is at the end of a pier near Chrissy Field in San Francisco; it has been recording water levels since June 30, 1854.)

At first surveyors used tide staffs, which were basically tall wooden rulers that had to be read by a person on the spot. By the late nineteenth century, the staffs gave way to rudimentary tide gauges that consisted of a pipe mounted on the end of a pier with a float in it; the float was attached to a pen that would record the ups and downs of the water on a roll of paper. Today, tide gauges are extremely high-tech, using microwaves to measure the precise distance to the surface of the water, then beaming it up to satellites, making it instantly available to researchers around the world.

No matter how accurate tide gauges are, however, the problem is that their measurements are always relative to the land they are placed on—and that land is often moving. In some spots, like the Gulf Coast of Louisiana, the ground is sinking, because of subsidence from groundwater pumping or other issues, so if you look at just the tide gauge, it looks like the sea is much higher than in other places. Sometimes, in places like Alaska or Finland, the land is actually rising, due to a phenomenon called glacial rebound.

New York City is a good example of glacial rebound in action: twenty thousand years ago, during the last ice age, the weight of the ice sheets depressed ground beneath them, mostly in Canada and the upper United States, causing the ground in what is now the New York City area to bulge out (think about how a couch cushion bulges when you put your hand on it). That bulge under New York is now subsiding, causing land to sink, which increases the local rate of sea-level rise.

The solution to all these local variations, of course, was to average tide gauges around the world. Still, tide gauges could only offer a rough approximation of something as complex as global sea levels. New technology provided a better way. In 1992, NASA and CNES (the French space agency) launched TOPEX/Poseidon, the first satellite capable of precisely measuring sea-level change. Three more have been deployed in succession since then, each overlapping with the previous one to provide an uninterrupted twenty-five-year record of sea-level change. The most recent satellite is the Jason-3, which launched in early 2016. The Jason-3 circles the Earth continuously, bouncing radar waves off the surface of the sea to measure the distance between the satellite and the water, as well as the height of the satellite relative to the center of the Earth. These measurements, from which the influence of the tides and the waves is removed, are free from distortion by rising or sinking land. When this data is combined with tide gauge averages, as well as measurements from ocean floats that record changes in the heat content of the ocean, it gives scientists a very good picture of how much the sea level is rising and what the causes are.

With better data, scientists are now able to more clearly understand other factors beyond land movement that lead to variations in the rate of sea-level rise. One is the gravitational fingerprinting I mentioned earlier, which pushes water into the Southern Hemisphere from melting ice sheets in Greenland and into the Northern Hemisphere from Antarctica. Another important factor is temperature, which fluctuates daily, seasonally, and annually. As water heats up, it expands (eventually it boils and turns to vapor, of course—not something we have to worry about with the world's oceans anytime soon). Globally, the thermal expansion of the oceans caused by the Earth's rising temperature has contributed about half of the observed sea-level rise in the last fifty years. In the future, that percentage will decline as thermal expansion is dwarfed by increasing melt rates in Greenland and Antarctica.

Ocean currents also impact regional sea levels. On the East Coast of the United States, the speed of the Gulf Stream—the underwater current that carries cold water from the north down to the equator, then loops around and carries warm water back up to the Arctic—can influence sea levels all the way from Virginia to Florida. A faster Gulf Stream pulls water away from the coast; as it slows down, it walls up, raising sea levels. Norfolk, Virginia, which is a sea-level-rise hotspot due to its low-lying topography and ground subsidence, has not been helped by the fact that a slowing Gulf Stream is pushing more water against the coast. One recent study showed that, between 1950 and 2009, the seas north of Cape Hatteras rose three to four times faster than the global average.

The most surreal consequence of melting ice and rising seas is that together they are a kind of time machine, so real that they are altering the length of our days. It works like this: As the glaciers melt and the seas rise, gravity forces more water toward the equator. This changes the shape of the Earth ever so slightly, making it fatter around the middle, which in turns slows the rotation of the planet similarly to the way a ballet dancer slows her spin by spreading out her arms. The slowdown isn't much, just a few thousandths of a second each year, but like the barely noticeable jump of rising seas every year, it adds up. When dinosaurs roamed the Earth, a day lasted only about twenty-three hours.

*   *   *

Jason Box was born in Colorado and spent his early years in suburban Denver, where his father worked as an electrical engineer for an aerospace company. "Jason was smart, and

Jason Box coring ice in Greenland. *(Photo courtesy of Peter Sinclair)*

he liked to cause trouble," recalls Box's older sister Leslie. When Box was about ten, he erected a lightning rod in a field and nearly burned the town down.

As a teenager, Box often wore a sleeveless military jacket and blasted around on a skateboard listening to bands like the Dead Kennedys, Bad Religion, and Judas Priest. He enrolled at the University of Colorado Boulder, where he and his sister played in a garage band called the Sensors (Box played guitar and sang). During gigs, Box liked to decorate the stage with old electronic equipment like oscilloscopes and broken fax machines. "Jason was one of those guys who could drink hard and party at night and then get up early in the morning and go to an intense science class," his sister recalled. At Boulder, he bounced from computer science ("too nerdy") to astronomy ("amazing but not down-to-earth") to geology ("too slow"). "Then I took a climatology course and saw the Keeling curve and that did it," Box told me. The Keeling Curve, named after scientist Charles David Keeling, is the famous graph that measures the rise of carbon dioxide concentrations in the air since 1958, which is the bedrock of global warming science. "I knew the implications were huge," said Box.

Box took his first trip to Greenland when he was twenty years old. "I remember him walking into my office and saying, 'I want to go to Greenland with you,'" said Konrad Steffen, a top glaciologist who at that time was a professor of geology at the University of Colorado. Box worked to install and maintain mini–weather stations around Greenland and became increasingly interested in how wind, temperature, and sunlight affect glaciers. After writing a PhD thesis on how much Greenland glacier ice is lost from evaporation, Box took a position at Ohio State University, home of the prestigious Byrd Polar and Climate Research Center. In

2013, Box, his wife, Klara, and his young daughter, Astrid, moved to Copenhagen and joined the Geological Survey of Denmark and Greenland. One of Box's first tasks was to perfect a computer model that monitors Greenland's mass ice change in real time; he was also involved in helping the Danish government review projects in Greenland that might exploit the melting ice, such as new hydroelectric dams.

In the summer of 2012, Box was at New York's LaGuardia Airport on his way to Greenland when he saw the first video images of the massive wildfires in Colorado. "It's a strange feeling, watching your home state burn on TV," Box told me. But it gave him an idea. NASA's Thomas Painter recalled, "Jason called me and said, 'Do you think soot from wildfires might be melting Greenland?' I told him that I didn't know if soot particles were landing there, but it was certainly conceivable, given the circulatory patterns of the Earth's atmosphere." A few weeks later, the case for this grew stronger when Box was scanning laser satellite images of Greenland and discovered a cloud of smoke—possibly drifting soot from a wildfire—over the ice.

The idea that soot can have a powerful effect on the melt rate of snow and ice is not new. NASA scientist James Hansen explored the idea in a paper published in 2004, arguing that if soot reduced the reflectivity of Arctic ice by just 2 percent, it had the same effect on the melt rate of the glacier as a doubling of CO_2 concentrations in the atmosphere. What was new was Box's attempt to link the Colorado wildfires directly with the 2012 meltdown in Greenland—to make a direct connection between a particular fire and a particular melting event.

A lot of scientific ideas are poetic, but this one really underscores the way small changes in the climate can amplify each other in unpredictable ways. As Box sees it, warmer temperatures in the United States stress pine trees in the Rockies, leaving them vulnerable to pine bark beetles, which bore into the trunks of the heat-weakened trees, killing them and turning them into tinder. A backpacker's campfire throws out a spark, a tree ignites, and soon the mountainside is burning and the soot is drifting up, some of it lofted into the jet stream and settling in Greenland, darkening the snow and accelerating the transformation of ice into water, which runs down into the North Atlantic, and, eventually, pushes a little deeper into Miami, Shanghai, New York City, Venice, Mumbai, Lagos, and the rice fields of Bangladesh, and then a little deeper still.

Eugene Domack, a geologist at the University of South Florida, was one of the last human beings to see the Larsen B ice shelf on the Antarctic Peninsula. In December 2002, in the middle of the polar summer, he spent three weeks in Antarctica sampling mud from the bottom of the Amundsen Sea—he was looking for rocks that had been rafted out to sea on the bottom of icebergs, which would tell him something about how fast the ice sheets had broken up in the past. That December, he spent a lot of time in the waters just in front of the face of Larsen B. At the time, the Larsen B was one of the biggest ice shelves in the world—roughly the size of Rhode Island, and up to two hundred feet thick. That summer, Domack noticed it was particularly warm, and if he could have gotten up onto the top of the ice shelf, he would have seen melt ponds everywhere (the Antarctic Peninsula, which juts north off the continent, is the only part of Antarctica that has shown significant surface warming in recent decades). But Domack, who never got up on top of the ice, had no sense that the ice shelf was unstable. It had, after all, been there for twelve thousand years.

A month or so after he got back to the United States, he was astonished when Larsen B made international news: the entire ice shelf had collapsed in spectacular fashion. Its demise was chronicled by satellite images. The crack-up took less than a month. "It was mind-blowing," Domack told me. "No one thought it could disintegrate that fast."

For climate scientists, both the collapse of Larsen B and the full melt of the surface of Greenland's ice sheet a decade later were big wake-up calls. "It showed us how much we didn't understand about what is going on with the ice sheets," said Peter Clark, a leading sea-level rise expert at Oregon State University. Clark was one of the lead authors on the sea-level rise section of the fifth (and most recent) report by the Intergovernmental Panel on Climate Change, which was published in 2013—too soon for studies that explored the big melt in Greenland the year before to be included, as well as more recent studies that highlighted the fragility of West Antarctic glaciers. As a result, as soon as it was published, the 2013 IPCC report was already out of date.

This matters a lot, because the IPCC reports are important documents, providing the scientific basis for global climate agreements and coastal planning around the world. The 2013 IPCC report, which projected a high end of possible sea-level rise of about 3 feet 2 inches, was particularly important, because it was the scientific basis for the 2015 climate treaty negotiations in Paris, which were viewed by many politicians and activists as the last good shot to get a meaningful global agreement to reduce carbon pollution. The IPCC report, which gets updated about every six years, is a synthesis of both historical data (how fast and how high sea level has risen in the past) and state-of-the-art modeling and research. Of course, science moves slowly and deliberately, and the IPCC report does not sell itself as cutting-edge knowledge. Still, because the IPCC report is viewed as the gold standard by politicians and scientists, many people think the IPCC's 3-foot-2-inch high-end estimate for sea-level rise by 2100 is as bad as it can get. It's not.

After 2012, with the combination of the big melt in Greenland and the collapse of Larsen B, it was pretty obvious that the ice sheets were changing far faster than forecast. How much did scientists really understand now about what might happen in the future? NASA's James Hansen published a paper in 2015 stating that, due to the exponential increase in ice sheet melting in Antarctica, we could see as much as nine feet of sea-level rise by 2100. Another paper by Rob DeConto at the University of Massachusetts and David Pollard at Pennsylvania State University suggested that the rapid calving and retreat of big glaciers like Thwaites and Pine Island in West Antarctica could alone contribute over three feet of sea-level rise by 2100. And what scientists in the field were seeing with the breakup of ice shelves and the speeded-up melting of glaciers confirmed this. "The latest field data out of West Antarctica is kind of an OMG thing," Margaret Davidson, head of coastal planning for the National Oceanic and Atmospheric Administration, said in a 2016 email.

For anyone living in Miami Beach or South Brooklyn or Boston's Back Bay or any other low-lying coastal neighborhood, the difference between three feet of sea level rise by 2100 and six feet is the difference between a wet but livable city and a submerged city—billions of dollars' worth of coastal real estate, not to mention the lives of the 145 million people who live fewer than three feet above sea level, many of them in poor nations like Bangladesh or Indonesia. The difference between three feet and six feet is the difference between a manageable coastal crisis and a decades-long refugee disaster. For many Pacific island nations, it is the difference between survival and extinction.

The big question remained: What were the key drivers of rapid ice melt that scientists didn't understand or incorporate into their models? Was it a wobbling jet stream? Changing ocean currents? Soot on snow? In their paper on West Antarctica, DeConto and Pollard had gotten much faster collapse of the glaciers simply by factoring in the effect of a small amount of meltwater on the surface of the West Antarctic glaciers, as well as a better understanding of the physics of ice-cliff fracturing. "We just don't know what the upper boundary is for how fast this can happen," Richard Alley, a geologist at Penn State University who probably understands ice sheet dynamics better than anyone, told me. "We are dealing with an event that no human has ever witnessed before. We have no analogue for this."

Box, like most scientists I've talked to, readily acknowledged that the IPCC estimate of 3 feet 2 inches of sea-level rise by 2100 is far too low. I asked him if he thought 6 feet by 2100 was still too low.

Box replied without hesitation: "Shit yeah."

Finally, dinners consumed and nervous feet nearly tapped out, Box was able to find the helicopter pilot and we were ready to go. We hopped on a short flight to Ilulissat, which was about a hundred miles up the western coast of Greenland. As soon as we got into the air, I could see the ice sheets—they looked like big white rivers flowing down to the dark blue sea. As we neared Ilulissat, Box pointed out the Jakobshavn to me—we went right over it. You could see the wide, flat catchment area of the glacier, and the river of ice that fed down toward Disko Bay, abruptly stopping at the water's edge. Icebergs were scattered in the fjord in front of it. From the air, they looked like pebbles.

We landed at the small airport at Ilulissat, unloaded our gear, and took a shuttle bus to our hotel. Ilulissat (Greenlandic for "icebergs") was charming, a fishing village with brightly colored New England–style wooden houses sloping down to an ocean cove. The Hotel Arctic, where we were staying, was perched on the hillside above the cove with views out toward the fjord. It felt like the Waldorf compared to the hut we'd stayed at in Kangerlussuaq ("This is where Al Gore stays when he visits," I heard a guest at the Hotel Arctic say as we were checking in). While Box double-checked his gear, I went for a short walk along the bluff near the hotel. Icebergs drifted not far offshore, glinting in the afternoon light. Some were as big as the New York Public Library on Fifth Avenue, others as small as a mailbox. They reminded me of a battalion of soldiers heading into battle.

In the morning, we took a shuttle to the airport, where we met Malik Nielsen, our next helicopter pilot. He was by

Icebergs in Disko Bay, Greenland. *(Photo courtesy of the author)*

turns edgy and relaxed, a man who knew what he was doing but never forgot the thrill and danger of it. We discussed the flight plan, which was to fly out and have a look at the calving front of the Jakobshavn, then go up onto the ice field to collect soot samples from the surface of the ice sheet.

We climbed into our Bell 212 helicopter, a common workhorse in Greenland, which was surprisingly spacious inside. Peter Sinclair, a filmmaker who joined us for the trip, sat up front with the pilot; Box and I were in back. We slipped on our headsets, double-checked everything, and were off for the twenty-five-minute flight to Jakobshavn.

We flew right along the bay in front of the glacier at an altitude of about 500 feet. It was like flying over an iceberg spawning ground. The face of the Jakobshavn loomed ahead of us, a blue-white wall of ice. I watched chunks of ice calving off, falling into the sea.

Before we got to the face, the chopper turned south and flew toward the rough mountains alongside the glacier. We were looking for a patch of open ground that Box had identified in satellite photos. Box and the pilot exchanged a few words on the intercom; then Box smiled and gave me a thumbs-up. A few moments later, the chopper touched down on an unremarkable bit of rocky tundra about the size of a football field, and Box jumped out. "Welcome to New Climate Land," he said, then launched into a giddy, erudite stand-up monologue for Sinclair's camera that would have made his high school science teacher proud. For thousands of years, he explained, this spot had been covered by a tall building's worth of ice and snow. But now, in just the last few months, the last traces of that ice and snow had disappeared. "We are likely to be the first human beings to ever stand on this piece of ground," Box said excitedly.

Before we took off for the interior of the ice sheet, I asked if we could take another pass along the front of the Jakob-shavn. Box nodded but warned that we couldn't get too close. Big calving events are unpredictable and can create danger-ous air turbulence as massive ice cliffs collapse into the fjord.

Apparently Nielsen wasn't too worried about this, because as we approached the calving front of the glacier, he flew so close that I felt like I could reach out and touch it. A wall of ice streamed by the chopper window—blue, translucent, cracked. Just ahead of us, I watched a huge chunk collapse into the water. It fell straight down, like a trapdoor had opened beneath it. I looked again, and another smaller piece fell. I could feel the physics at work here, the ancient and unstoppable force of a glacier sliding down into the sea and reshaping our world.

4. AIR FORCE ONE

PRESIDENT OBAMA WAS in a very good mood in Alaska. It was the fall of 2015, and he had a little more than a year left in his second term—you could tell he had the finish line in sight. The stated purpose of the trip was to draw attention to the looming climate catastrophe the world faces, but with the exception of one big policy speech in Anchorage, in which he sounded as apocalyptic as any hemp-growing activist, he spent most of his three days up north beaming. "He's happy to be out of his cage," one advisor joked. Others credited the buoyant US economy or the fact that the president had just learned that he had secured enough Senate votes to protect the hard-fought and controversial nuclear deal with Iran. Donald Trump had announced his presidential run a few months earlier, but at this point, Trump's campaign was a joke that no one took seriously. President Obama's popularity was sky-high, and his legacy seemed secure.

Whatever the reason, you could see the cheerfulness in

Obama's face the moment he stepped out of his armored limo at Elmendorf Air Force base in Anchorage. The president was all smiles, shaking hands with local pols and then bounding up the stairs into Air Force One for the short flight to Kotzebue, a village on the west coast of Alaska that is threatened by sea-level rise and other climate impacts. No suit and tie, no sir—today, the third and final day of his Alaska trip, he was dressed for adventure in black outdoor pants, a gray pullover, and a black Carhartt jacket. As White House press releases and video blogs pointed out, this was a historic trip—not only would Obama be the first sitting president to visit the Arctic, but he would also be the first president to use a selfie stick to take videos of himself talking about the end of human civilization.

The president's upbeat mood was an odd and unexpected counterpoint to the seriousness and urgency of the message he was trying to deliver. "Climate change is no longer some far-off problem; it is happening here, it is happening now," he said in his remarks to an international summit on the Arctic in Anchorage on the first day of his trip. In surprisingly stark language, Obama warned that unless more was done to reduce carbon pollution, "we will condemn our children to a planet beyond their capacity to repair: Submerged countries. Abandoned cities. Fields no longer growing." His impatience was obvious: "We're not acting fast enough," he said four times in a twenty-four-minute speech (an aide later told me this repetition was ad-libbed).

For Obama, this trip to Alaska marked the beginning of the last big push of his presidency—to build momentum for a meaningful deal at the international climate talks in Paris later that year. ("I'm dragging the world behind me to Paris," Obama later told a visitor to the Oval Office.) The Paris agreement was widely viewed as a last-ditch effort to get the nations of the world to commit to reducing carbon pollution to a level that might limit the worst impacts of climate change, including slowing sea-level rise in the decades to come.

Policy-wise, the president didn't have much to offer in Alaska. He restored the original native Alaskan name to the highest mountain in North America (Denali) and accelerated the construction of a new US Coast Guard icebreaker—largely symbolic gestures that didn't do much to help Alaskans deal with eroding shorelines and thawing permafrost (he would later propose $100 million to relocate Alaska villages in his 2017 budget proposal, but the funding was redlined out). In the end, the trip was mostly a calculated and well-crafted presidential publicity stunt. And it raised the question: If the American people see the president of the United States standing atop a melting glacier and telling them the world is in trouble, will they care?

"Part of the reason why I wanted to take this trip was to start making it a little more visceral and to highlight for people that this is not a distant problem that we can keep putting off," the president told me. "This is something that we have to tackle right now."

Obama could not have picked a better place to make his point than Alaska. Climate-wise, it is the dark heart of the fossil fuel beast. On one hand, temperatures in the state are rising twice as fast as in the rest of the nation, and glaciers are retreating so quickly that even the pilot of my Delta flight into Anchorage told passengers to "look out the window at the glaciers on the left side of the aircraft—they won't be there for long!" And it wasn't just villages like Kotzebue that were in trouble. The very week of Obama's visit, thirty-five thousand stressed-out walruses huddled on the beach in northern Alaska because the sea ice they use as resting spots while hunting had melted away.

On the other hand, the state is almost entirely dependent on revenues from fossil fuel production, which, thanks to the low price of oil and exhausted oil and gas wells on the North Slope, was in free fall—the state was grappling with a $3.7-billion budget shortfall that year. Alaska governor Bill Walker had flown from Washington, DC, to Anchorage with the president at the beginning of his trip; according to one of the president's aides, Walker more or less pleaded with the president to open more federal lands to oil and gas drilling to boost state revenues. "Alaska is a banana republic," Bob Shavelson, executive director of Cook Inletkeeper, an environmental group in Alaska, told me. "The state has to pump oil or die."

For the flight up to Kotzebue, the air force left the president's 747 parked on the tarmac in Anchorage and switched to a smaller plane, a 757 (it was also dubbed Air Force One, which name applies to any airplane the president is flying in—the president's staff called it "mini–Air Force One"). Several members of the president's senior staff were along, including Susan Rice, his national security advisor.

Rice's presence on the trip was a reminder that a rapidly melting Arctic also has rapidly escalating national security implications. As the ice vanishes, there's a whole new ocean opening up—and one that contains 30 percent of the known natural gas reserves and 13 percent of the oil. Unlike Russia, the US is poorly equipped to operate up there, with only one heavy-duty icebreaker (the Russians have forty). And the Russians aren't the only ones with eyes on the Arctic—at the very moment we were flying toward Kotzebue, five Chinese warships were cruising in international waters below. Coincidence or power play? And a few hundred miles to the east, the Canadian military was engaged in Operation NANOOK, an annual large-scale military exercise that, according to the Canadian government, was "about demonstrating sovereignty over northern regions."

Before we crossed into the Arctic, we touched down in Dillingham, a small town on Bristol Bay that is the heart of the salmon fishery in Alaska. The presidential motorcade headed straight for the beach, where a couple of native Alaskan women had caught a few silver salmon in a net, which made another nice visual tableaux for the president's social media feed and gave him a chance to talk briefly about the importance of salmon to Alaska's economy. The funniest moment of the trip occurred when the president, who was wearing orange rubber gloves, held up a two-foot-long silver salmon that the fisherwoman had given him. The salmon, apparently a male and still very much alive, ejaculated on his shoes. Obama laughed, and the fisherwoman said something privately to him. The president laughed again and repeated her remark loudly enough for everyone to hear: "She says he's happy to see me."

Last stop, Kotzebue. On the way, the president asked the pilot to fly a little farther north and circle over the island of Kivalina so he could get a look at it. Like Kotzebue, Kivalina is a poster child for the havoc that climate change is wreaking on native Alaskan villages along the coast—in addition to rising seas that are eating away as much as sixty feet of shoreline each year, thawing permafrost is destabilizing the soil, causing houses to collapse into the sea. About four hundred people live on Kivalina, and they are in trouble—relocating the village to higher ground on the mainland will cost $100 million or so, which neither the state nor the federal government has been willing to pay for.

We descended in a long arc over the island, which is less than a mile off the Alaskan coast—from the air, it looked

The barrier island of Kivalina, where 400 Alaskans live, is being swallowed by the sea. *(Photo courtesy of Shutterstock)*

like an Arctic version of Miami Beach, a thin barrier island floating in a vast cold gray sea. Hope Hall, the White House videographer, rushed to the window with her camera to take a shot of the soon-to-be-sunken land, which was later used in one of the president's video messages that he posted on Facebook and other social media sites.

We touched down in Kotzebue at about 5 p.m. The president was greeted on the tarmac by Reggie Joule, the mayor of the Northwest Arctic Borough; then we climbed into our assigned vehicles in the motorcade for the short drive to the high school. We rolled by flimsy weather-beaten houses with American flags hanging in the windows and broken dogsleds in the front yards. You could sense the hardship of life in a place where it gets down to 100 degrees below zero (with wind chill) in the long, dark winters and where the nearest road to civilization is 450 miles away.

The motorcade pulled up at Kotzebue High School, a large metal building draped with banners welcoming the president, with snipers pacing on the roof. A thousand people crowded into the basketball gym, draped with the blue and gold of the Kotzebue Huskies. Obama gave a relaxed speech about climate change and the wonders of the far north, clearly enjoying the fact that history would remember him as the first sitting president to visit the Arctic. He said he was envious that Warren Harding spent two weeks here during a trip in 1923, but explained that he had to get back quickly because "I can't leave Congress alone that long."

When it was over, a White House aide guided me into a nearly empty classroom with a large round table in the center and two blue plastic chairs. Ice crystals made from blue construction paper hung from the ceiling and a Secret Service officer kept watch by the door. I chatted with Josh Earnest, the White House press secretary, who was along for the trip, and fiddled with my notes. Remarkably, the White House did not put any limits on the scope of my questions or ask to vet them in advance. Earnest told me I would have forty-five minutes alone with the president.

I heard footsteps in the hallway, and then the president walked in. He was easy, familiar, and if you didn't know he was president of the United States, you might say there was nothing intimidating about him. We shook hands and exchanged a few words about the flight; then he sat down in one of the plastic chairs and said, "Let's do it." We talked for more than an hour—during which the cheerfulness that had animated his public remarks on this trip dissipated. He spoke in measured tones, but with a seriousness that suggested that he believed—not unjustifiably—that the fate of human civilization was in his hands. Only near the end, when I asked if he felt any sadness about what we are losing in the world as a result of our rapidly changing climate, did he show any emotion—he averted his eyes for a moment and looked away, as if the knowledge of what's coming in the next few decades was almost too much to bear.

One of the first things I asked him about was drilling for oil in the Arctic, which was much in the news during our visit. If he took climate change so seriously, how could he allow this? The president quickly made the point that opening the Arctic for drilling was not his doing. "One of the things about being president is you're never starting from scratch," he said, not mentioning that it was George W. Bush who had cleared the way for drilling off the coast. He argued that no matter how urgent the science is on climate change, you have to take the politics slowly, especially in a fossil-fuel-dependent state like Alaska. "If I howl at the moon without being able to build a political consensus behind me, nothing's going to get done." He talked about the importance of pushing clean energy, which can lower energy costs and create jobs, "so that we're reducing what is perceived as a contradiction between economic development and saving the planet."

"Okay, I understand that," I argued, looking the president in the eye, while simultaneously aware in another part of my brain that I was looking the president of the United States in the eye. "But the problem is that building consensus on climate change is different than other issues because you have physics to account for too, right? The warming of the planet is not waiting for consensus-building."

"I understand," the president replied coolly. He pushed his sleeves up, revealing his thin wrists. "But if we're going to get our arms around this problem, which I think we can, then we are going to have to take into account the fact that the average American right now, even if they've gotten past climate denial, is still much more concerned about gas prices, getting back and forth from work, than they are about the climate changing. And if we are not strategic about how we talk about the issue and work with all the various stakeholders on this issue, then what will happen is that this will be demagogued and we will find ourselves in a place where we actually have slower progress rather than faster progress.

"So the science doesn't change," he continued. "The urgency doesn't change. But part of my job is to figure out what's my fastest way to get from point A to point B—what's the best way for us to get to a point where we've got a clean-energy economy. And somebody who is not involved in politics may say, well, the shortest line between two points is just a straight line; let's just go straight to it. Well, unfortunately, in a democracy, I may have to zig and zag occasionally, and take into account very real concerns and interests."

I thought, "Obama the pragmatist." But given that he was now head of the world's biggest economy and leader of the free world, how did he handle the responsibility of avoiding a potential climate catastrophe within his daughters' lifetimes?

I thought he might balk at the phrase "climate catastrophe." But he did not.

"I think about it a lot," he said, pausing and looking down at his hands. "I think about Malia and Sasha a lot. I think about their children a lot."

Then, switching back to a more presidential-sounding voice, he went on: "One of the great things about being president is you travel a lot and you get to see the world's wonders from a vantage point that very few people get a chance to see. When we were out on the waters yesterday, going around those fjords, and the sea otter was swimming on its back and feeding off its belly, and a porpoise jumps out of the water, and a whale sprays—I thought to myself, 'I want to make sure my grandchildren see this.' "

President Obama on the beach in Dillingham, Alaska. *(Photo courtesy of The White House/Pete Souza)*

"We go back to Hawaii every year, and I intend to, hopefully, spend a lot of time there when I'm out of office. I want to make sure my kids, when they go snorkeling, are seeing the same things that I saw when I went snorkeling when I was five years old, or eight years old. I spent a big chunk of my life in Indonesia when I was young, and I want them to be able to have some of the same experiences, walking through a forest and suddenly seeing an ancient temple. And I don't want that gone."

The president mentioned that during his vacation, he had read *The Sixth Extinction,* by Elizabeth Kolbert, which is about the impact climate change is having on the natural world. "It makes very clear that big, abrupt changes can happen; they're not outside the realm of possibility," he said. "They have happened before; they can happen again."

He made a little tent with the fingers of both hands, then continued. "So all of this makes me feel that I have to tackle this every way that I can. But one of the things about being president is you're also mindful that, despite the office, you don't do things alone. I continually go back to the notion that the American people have to feel the same urgency that I do. And it's understandable that they don't, because the science right now feels abstract to people. It will feel less abstract with each successive year. I suspect that the record wildfires that we're seeing, the fact that half of the West is in extreme or severe drought right now, is making people understand this better. If you talk to people in Washington State right now, I suspect, after having tragically lost three firefighters, and seeing vast parts of their state aflame, that they understand it better. If you go down to Florida, and neighborhoods that are now flooding just every time the tide rises, they're understanding it better."

After the formal interview, the president left to meet with some local officials. When he was finished with that, he and I were scheduled to take a walk along Kotzebue Sound so that we could be filmed together for a short documentary *Rolling Stone* was producing about the visit. We rode in the motorcade a few blocks to the water's edge. The water of Kotzebue Sound was gray and flat, and even though it was only early September, you could already feel winter approaching. A few hundred feet off, the president stepped away to talk to locals about vanishing sea ice and flooding.

I thought back to a few days earlier, when I'd seen the president walk up to the face of a mountain glacier. It was not a big glacier, but even a small glacier made the president look diminutive. On this trip I'd witnessed all the trappings of presidential power—the jets, the helicopters, the Secret Service agents, the obsequiousness of local politicians. But compared to the larger forces at work in the natural world, it was nothing.

After about ten minutes, one of the president's aides waved me over to join them, and the president and I took a walk along the newly armored shore of Kotzebue, where tons of riprap and concrete protected the town from the rising waters of the sound. Secret Service agents trailed behind, just out of earshot. It was a little chilly—the president had his hands in his coat pockets.

"I've talked to a lot of scientists about climate change," I said, not wanting to waste any time. "A lot of them wrestle with how honest to be about what they see coming—how blunt, how optimistic. You obviously have a great responsibility on this—how do you gauge how much truth America can take? Because you know what's coming..."

"Well, here's the thing," he told me, looking out over the sound as we walked. "When I was a community organizer, one of our basic principles was: if you have a big problem, you have to break it down into pieces that people can absorb. So if you talk to people about world hunger—their general attitude is, well, 'I can't solve world hunger.' If you talk to people about 'Let's solve this particular problem that alleviates hunger for these kids,' then you can get some action. So my job up here, the whole point of this trip, is to sound the alarm. But I want to make sure that I'm not presenting this in a way that leads people to think that we're doomed and there is nothing we can do about it."

The way he said this, I wasn't entirely convinced he believed we weren't doomed. But I let it pass.

"Look, we're not going to save every frog and we're not going to save every coral reef," he said. "But I can save some coral reefs, and I can save some frogs. And there's gonna be adaptations that have to be made and there are gonna be displacements, but I can mitigate the worst and hope that the planet ends up being more resilient than it may feel to be right now. But the most important thing is at least to express urgency and not hold back from the fact that we need to be a lot more aggressive than we are right now."

I pressed on. "But you know the science—doesn't it scare the hell out of you sometimes?"

"Yeah," he said, simply and flatly.

I brought up the fact that many scientists believed that we could see six feet or more of sea-level rise by the end of the century, which was twice the IPCC estimates.

"Six feet?" the president said, as if hearing the words suddenly made the idea all too real to him.

"Yeah," I said. "As you know, there is some uncertainty in these studies, but the error bars are all in the direction of more sea-level rise than we anticipate, not less...."

"Look, part of my job is to read stuff that terrifies me all the time."

I couldn't help but laugh, the way he said it. "That's true, I suppose."

"I've got a chronic concern about pandemics, for example. And the odds are that sometime in our lifetime there's gonna be something like the Spanish flu that wipes out a lot of people...if we're not taking care. I do what I can do and as much as I can do and what I don't want to do is get paralyzed by the magnitude of the thing and what I don't want is for people to get paralyzed thinking that somehow this is out of our control. And I'm a big believer that the human imagination can solve problems. We don't usually solve them as fast as we need to. It's sort of like two cheers for democracy. It's that kind of thing. We try everything else, I think Churchill said, eventually, and when we've exhausted every other alternative we finally do the right thing. Hopefully the same will be true here."

With that, we came to the end of our walk, and one of the president's aides guided him over to meet with 2011 Iditarod champion John Baker, who gave him a sled dog puppy to hold and a baseball cap to take home. While I watched, I talked with Brian Deese, special advisor to the president on issues like climate change. He was a smart and unpretentious guy, secure enough in his standing with the president that he could wear ragged hiking shoes with a sole falling off and think it was kinda funny.

I told Deese that during our talk, the president had mentioned the risks Florida faces from sea-level rise. I wondered how people like Deese, who was one of the prime architects of the president's successful auto industry bailout in 2009, thought about the looming financial disaster that sea-level rise would bring to coastal cities. "What are you going to do when Miami goes under?" I asked him. "What does it mean to lose a great American city? How is the federal government going to deal with that? If Miami needs money, so will a lot of other places along the coast."

"Miami has lots of resources," Deese told me. "They can figure this out. They might have to raise taxes to pay for it, but I think you'll see a lot of innovation down there in the coming years. I'm more concerned with places like Kivalina, where people have nowhere to go and very few resources for themselves."

A moment later, Deese left to join the president, who was

heading back to his armored SUV. At about 8:30 p.m., we motorcaded back to the airport and the president bounded up the steps to Air Force One. A small group of Alaskans waved at him from behind a chain link fence and shouted goodbyes. He had been in the Arctic for about four hours— but that was four hours more than any other president had committed. As I took my seat on Air Force One, I noticed that the president was already seated in his leather chair at the conference table on the plane, still wearing his Iditarod hat. He said to his staff, "Let's get to work."

5. REAL ESTATE ROULETTE

THE PÉREZ ART Museum Miami, which opened in 2013, was the first building in the city to acknowledge that Miami is a new Atlantis—even if it did so obliquely and unintentionally. Designed by the renowned Swiss firm Herzog & de Meuron, the museum, which cost $118 million to build, sits at the edge of downtown Miami, overlooking Biscayne Bay. The architects were inspired by Stiltsville, a collection of houses built on stilts in the bay in the 1930s, largely to give owners a place to party that was beyond the reach of the law. The museum seems to be suspended in midair, with the main structure supported by a series of thin concrete pillars. A flat pergola-like roof lets sunlight filter through, and a big deck overlooks the water, with hanging gardens dangling like seaweed. The ground floor is an open gravel parking lot—you can imagine a storm surge pushing through, then receding without much damage beyond a few banged-up cars. The entire building seems to hover at the edge of the bay, as if it were waiting for the water to come. The museum, which has been widely praised as a landmark in the cultural evolution of Miami, bears the name of developer Jorge Pérez ("the Trump of the Tropics," *Time* dubbed him), who contributed $55 million in cash and artworks in exchange for the right to put his name on the building.

I visited the museum one night to hear a talk by artist Michele Oka Doner, who had a retrospective at the museum called *How I Caught a Swallow in Midair*. Oka Doner grew up in Miami Beach, where her father was a judge and later, in the 1950s, the mayor. Her work is deeply connected to Miami Beach, much of it inspired by years of walking along the shore: sargasso weed shaped into human figures, bronze casts of coral assembled into chairs, driftwood sculptures that suggest mythical sea creatures. Oka Doner is best known for her public art installations around Miami, including a piece called *A Walk on the Beach,* which covers more than a mile of the floor in Miami International Airport, where black terrazzo is embedded with bronze objects shaped like seashells and diatoms. I had walked over it many times before it occurred to me that it felt less like a walk on the beach than a walk underwater. As if she were suggesting that Miami airport—which is located in a particularly low-lying part of Miami-Dade County and floods regularly—had already been reclaimed by the sea. How subversive! When I mentioned this to Oka Doner, she said, with a twinkle in her eye, "Well, yes, that was what I intended."

During the museum talk, she reflected on her work and how she was inspired by the natural world of South Florida. She didn't say anything to this well-dressed crowd of Miami power brokers and art collectors about sea-level rise or climate change, despite the fact that she understood very well what was at stake—she had recently sold her home on Miami

Beach in part because she believed it was time to get out before the waters rose.

When the talk was over, I wandered into the lobby and stood in line to purchase a catalogue of her show. In front of me, I noticed a man in a nice charcoal-gray suit. When he turned around, I recognized him immediately: Jorge Pérez, the namesake of the museum.

If any one person symbolizes the early-twenty-first-century building boom in Miami, it is Pérez. "He's the eight-hundred-pound gorilla of Miami developers," condo analyst Peter Zalewski told me. Pérez, who was sixty-six when we met at the museum, is a son of Cuban refugees who was born in Argentina, raised in Colombia, and educated at the University of Michigan. After college, he spent a few years in Miami's city planning department, then got into the development business. Related Group of Florida, which he cofounded with New York developer Stephen Ross, is now the biggest builder in Miami—one out of every five condos in the city has been built by Related. Related's buildings are known among architects for their unremarkable design and big profit-margins. ("He dares to dream, and to make those dreams real," his pal Donald Trump wrote in an introduction to *Powerhouse Principles,* Pérez's book about how to get rich in the real estate market. "The result is changed lives, and cities.") Pérez is an influential figure in Florida Democratic politics and gave generously to both Clinton's and Obama's presidential campaigns. In 2017, *Forbes* estimated his personal wealth at $2.8 billion.

While reporting this book, I tried to contact Pérez several times through his office to discuss sea-level rise and its impact on his buildings, but I had no luck. Now here he was, his suit crisply tailored, his tie beautifully knotted, his face tan and inexpressive, his moustache and beard neatly trimmed,

Jorge Pérez with Donald Trump in Sunny Isles, Florida, in 2005. *(Photo courtesy of Getty Images)*

his dark hair impeccably combed. A controlled man, used to keeping his cool.

I introduced myself as a journalist who was working on a book about sea-level rise. His face hardened. We chitchatted a bit. He mentioned that he had been a collector of Oka Doner's art for twenty years.

"I can't pass up this opportunity to ask you a few questions," I said to him. His face turned stony. "How is sea-level rise changing your thinking about the real estate business in South Florida?"

"We don't think about it on a daily basis," he replied.

I was surprised he was so dismissive.

"Does it change your thinking about the kind of property you want to develop?" I asked.

"No, it doesn't."

"Does it change the design of buildings you're building?"

"No," he said, beginning to sound agitated. "We build to the building code."

"Did it influence the design of this museum?"

"That is not something I gave any thought to," he replied.

"Well, aren't you worried that increased flooding in the city will impact the value of your real estate holdings? I mean, it's inevitable, isn't it?"

"No, I am not worried about that," he said. "I believe that in twenty or thirty years, someone is going to find a solution for this. If it is a problem for Miami, it will also be a problem for New York and Boston—so where are people going to go?" He hesitated for a moment, then added: "Besides, by that time, I'll be dead, so what does it matter?"

I wasn't sure how to respond. I understood that he wasn't thrilled about being questioned by a journalist at a social event. Nevertheless, his unwillingness to even nod to his responsibility for shaping the Miami of the future, or for the legacy of soon-to-be-swamped buildings that he was leaving behind, was startling. It is important to note that this encounter occurred in the spring of 2016, a time when Miami was already suffering sunny-day flooding at high tide and many major newspapers, magazines, and news programs were talking about the risks of sea-level rise in the city. It was impossible to imagine that Pérez didn't think about this or consider how it might impact his business. But it was very possible to imagine, as a Miami architect who has worked with Pérez suggested to me, that Pérez was afraid that if he talked about sea-level rise or acknowledged the risk, it would call into question some of his waterfront projects and, ultimately, cost him money.

In any case, I didn't have the opportunity to press him any further. We had by then inched to the front of the line, where Oka Doner was signing books. He turned away from me and said, "That was a wonderful talk, Michele."

She greeted him warmly and signed his book with a theatrical scrawl.

More than three quarters of the population of Florida lives on the coast, where virtually every house, road, office tower, condo building, electrical line, water line, and sewer pipe is vulnerable to storm surges and high tides. As the seas rise in the coming years, the vast majority of that infrastructure will have to be rebuilt or removed. According to a report by the Risky Business Project, a group cofounded by billionaires Michael Bloomberg, Tom Steyer, and Henry Paulson, between $15 billion and $23 billion worth of Florida real estate will likely be underwater by 2050; by 2100, the value of the drowned property could go as high as $680 billion.

In Miami, an awareness of what's to come is slowly percolating through the consciousness of property owners and real estate investors. For people who own a house or a condo, it mostly takes the form of a question: Should I sell or not? Can I get a few more years out of this place, or should I dump my condo on the beach now? Virtually everyone I know who owns property in Miami makes this calculation. It's like a game of real estate roulette — how lucky do I feel? How big a bet do I want to make?

Everyone plays the game in a different way, based on some combination of rumor, science, instinct, emotional connection to where they live, and tolerance for risk. The night I attended the talk at the museum by Oka Doner (who, as I mentioned, had made her own calculation and decided to sell her place in Miami Beach), my Uber driver turned out to be keenly aware of the risks of sea-level rise. Kamel had emigrated to Florida from Turkey a decade ago. Now he owned several condos in Miami—"I rent them out on Airbnb," he told me. When I asked him about sea-level rise, he didn't dispute that it was happening—in fact, he mentioned reading a magazine article that said the city had until 2025 before it got really bad. "So I have another seven or eight years before I have to sell," he said. "As long as people keep coming, I can make money on Airbnb."

A few nights later, I had dinner with a wealthy retired businessman who owned a spacious condo on the seventeenth floor of a building in one of the most flood-prone neighborhoods of Miami Beach. On a warm evening, we stood on his balcony and looked out over the waters of Biscayne Bay and the lights of downtown Miami. "I love this place," he said. "I feel so lucky to be here." He looked at the yachts below. "I think it is all defensible until we get to three feet," he said. "Even after that, I think we'll be okay here—this building is worth too much to just let go. Something will be done about it." But then he paused and pointed toward North Miami Beach, where property values are considerably lower. "But if you have an ordinary house on a lot by the golf course or something, I think you're in trouble. Nobody will care about that." A few weeks later, a friend who lives in the same building emailed me to tell me he had come to a different judgment. "I'm getting out while I still can," he told me. "The party is over." He had purchased his condo five years earlier for a million dollars. He ended up selling it for two million.

Everyone has their own strategy for real estate roulette. I met an artist who predicted that the Miami real estate market would tank when salt water upwelling through the limestone reached the roots of the trees and they started to die, which, in this artist's view, would be a powerful visual sign of the coming inundation. A schoolteacher told me that she and her husband were planning to sell when the value of their $350,000 house dropped to $300,000 on Zillow. And I met more than a few people who told me they would never sell, that they loved their lives in Miami too much and were going down with the ship.

Swiss psychiatrist Elisabeth Kübler-Ross wrote about the five stages of grief: denial, anger, bargaining, depression, and acceptance. As far as I can tell, until about 2013, there were only about four people in all of Miami-Dade County who would openly admit that sea-level rise was a serious issue for the city in the near term. As recently as 2010, when the county finalized a new zoning plan called Miami 21, which was supposed to celebrate the values of New Urbanism and prepare Miami for the twenty-first century, sea-level rise wasn't even mentioned. As one Miami-Dade County commissioner told me, "People thought that if they ignored the problem, it would go away."

But that denial is fading. Many Floridians have skipped over the anger stage and moved on to bargaining—especially over real estate. The issue for most people I talked with was not whether Miami will someday be underwater—it surely will be. The more pressing question was, How long should I stick around?

If there is such a thing as the Miami Beach sea-level rise intelligentsia, they were all gathered one night in early 2016 in the banquet room at the W, one of the swankiest hotels on the beach. Officially, the event was hosted by Miami Beach Chamber of Commerce, and the theme of the evening as stated on the program was the Economic Impact of Sea Level Rise. The unstated theme of the evening was *Holy shit, this is real—what are we going to do about it?*

During cocktail hour, I talked with Thomas Ruppert, a lawyer who works on coastal planning with Florida Sea Grant, a nonprofit group that works with governments and universities. "Sea-level rise is like aging," he told me. "You can't stop it, you can only do it better or worse." He talked about a guy in Key Largo who turned in his Bentley for a pickup truck because he was worried about saltwater corrosion, and about the complex legal issues surrounding the question of whether you still own your property when it goes underwater (in most cases, the answer is no). I chatted with a real estate broker who was apoplectic over a talk she'd heard that afternoon about whether real estate brokers should be required to disclose flood risks related to sea-level rise on properties they sell. "That would be idiotic," she told me, gulping down a gin and tonic. "It would just *kill* the market."

When the cocktail party ended, we moved into a banquet room, where a dozen or so tables were set up. The idea was that eight people would sit at each table, including two "influencers," who would each make a brief presentation and then invite discussion. After fifteen minutes, the influencers would move on to the next table, and so on.

By chance, I was seated at a table with Miami developer David Martin, who was head of Terra Group, a boutique firm that specializes in high-end developments, including a pair of towers in Coconut Grove designed by architects like Danish *enfant terrible* Bjarke Ingels. Martin, who is Cuban-American, is in his early forties but still has the energy of a teenager. He was born and raised in the Miami area, and by all accounts, had a deep feeling for the place. At the W, he wore his dark hair swept back with gel, stylish black-rimmed glasses, and white jeans and a tight white dress shirt. Architect Reinaldo Borges later described Martin to me as "a developer with a conscience."

The first speaker at our table was University of Miami geologist Hal Wanless, whom I had gotten to know several years earlier. As always, he looked like he'd just come in from a geology expedition, wearing a white short-sleeved shirt with a rumpled jacket thrown over it. He distributed a six-page handout called "The Coming Reality of Sea Level Rise—Too Fast Too Soon." He said to all of us at the table, "First thing you need to know is that global warming is real." He talked about how half the warming since 1997 has been stored in the ocean, which meant that even if we slowed CO_2 pollution now, the climate would keep heating up for a long time. He talked about how the city of Coral Gables is developing a plan to deal with sea-level rise in six-inch increments. "That's the kind of smart planning we need," Wanless said.

"Okay, I understand that," David Martin said. "But what I want to know is, how much sea level are we going to see—and how fast?"

"Well, consensus right now is three feet by 2100," Wanless replied. "But that keeps going up. I don't think it will be less than four feet by the end of the century. I personally believe it will be fifteen feet."

There was a moment of silence at the table. Eyes widened.

An expensively dressed real estate broker who was seated near me challenged Wanless: "This can't be a fear-fest!" she protested, sounding like a six-year-old on the verge of a temper tantrum. "Why is everyone picking on Miami? Why have we become the poster child on this? It is happening all over the East Coast, and the media is picking on Miami."

"Well, Miami has a lot at stake," Wanless said.

"I think scientists are pushing this because they want money," the real estate broker argued.

"I don't want money," Wanless said, indignant.

At left, Miami today. At right, Miami with seven feet of sea-level rise. *(Illustration courtesy of Climate Central/Hiasam Hussein)*

"Maybe not you personally, but you want it for development, for your science program at the university. You can't scare people, you can't tell them that Miami is not going to exist. It's not right. It's not fair."

"I'm just telling you what the science says," Wanless deadpanned. Then it was time for him to rotate to a new table.

Our second speakers came to the table—one of them was Miami lawyer and climate change advocate Wayne Pathman, who had organized the event. "Sea-level rise is a gamechanger," he told us. He talked about the importance of thinking ahead, especially on big building projects. As an example of how *not* to do it, he mentioned the renovation of the Miami Beach convention center, a $600-million project that does not take sea-level rise into account. "Why would we spend all that money and not elevate it? We could have built water retention areas to help with the flooding, but we didn't—how crazy is that?" He also brought up the causeways linking Miami Beach with the rest of the city: "We have three bridges, and they're all vulnerable—with just one and a half feet of sea-level rise, you won't be able to get to Miami Beach."

There was much discussion about building codes and height restrictions on buildings. The speakers rotated again, and Josh Sawislak, global director of resilience at AECOM, a global engineering company that specializes in big infra-structure projects, arrived at our table. He was a big guy with glasses and a cheerful manner. Before joining AECOM, Saw-islak had been an assistant director of the White House Council on Environmental Quality. He made the case that Miami Beach drives the entire economy of South Florida, and that Florida depends on what happens here. "It's vital that we keep Miami Beach thriving."

"Yeah, it is, bro," David Martin said, looking up from the notes he had been scribbling. "We need a team of fifty peo-ple, living and breathing this, who can go away for a year and come up with a solution for this. We need to show the world that there *are* solutions." Martin pointed out that some roads are only five to eight feet above sea level, but that the new buildings he was putting up were fifteen to eighteen feet above. "How does that make sense?"

Nobody had a good explanation.

A moment later, Sawislak said, "Miami needs to rebrand itself as capital of resilience."

"Very true!" the real estate broker exclaimed.

Martin, in his blunt way, turned to Sawislak: "How much will it cost to fix the city?"

"I don't know," Sawislak said.

"Can you raise the city up?"

"Yes, but..."

"Can you do it?" Martin pressed.

"Yes, we can do it," Sawislak said.

"So how much will it cost? Give me a number. A billion? Five billion?" Martin was growing impatient. "I want to know what it would cost to fix the city," he insisted.

"Well, you have to start from the top, and then work down to local implementation...."

Martin was not satisfied. "The problem is, how do you make budget decisions, how do you allocate tax dollars, if you don't know what it costs?"

"Well, we could do a study, give you an estimate...."

"We need hard numbers, then we can get something done. I just want to know what it's going to cost."

"Well, it's a big number."

"I know. But we're not a Third World country."

"No, we are not," Sawislak said.

Shortly after that, the meeting broke up.

There are few corporate headquarters in South Florida, no manufacturing to speak of, no entertainment industry (except sports and porn). Even the illegal drug market, which powered the Miami economy in the 1970s and 1980s, has declined. The core business of Miami is real estate and tourism. It is an empire of property and pleasure.

That makes Miami especially poorly suited to deal with sea-level rise. You go to the beach to escape problems, not to submerge yourself in them. Away from the coasts, Miami is a poor, gritty city with a long history of racial conflict, but all that isn't part of the economic engine of the region, it's the exhaust fumes. Miami is about fantasy, about reimagining yourself. It's not about contemplating the moral implications

The mouth of the Miami River in downtown Miami. *(Photo courtesy of the author)*

of choosing to drive an SUV or worrying about water damage to your leather sofa.

A second problem, which is connected to the first, is that there is no state income tax in Florida. State and local governments are largely funded by property taxes. In Miami-Dade County, for example, about one third of the county's operating budget comes from property taxes, which means that property taxes are hugely important to keeping schools open and police officers happy. And for all intents and purposes, there are only two ways to raise property tax revenue: increase the tax rate or build more and more expensive real estate. In Florida, which prides itself on being a low-tax state, even talking about raising rates is a likely career-ender for any politician who suggests it. So the incentive here has been to keep building and building and not do anything to rock the boat or to make investors fear that the money they've parked in that oceanfront condo is not safe.

The third and biggest issue is, nobody wants to spend money to build a more resilient city because nobody owns the risk. When a developer like Jorge Pérez builds a new condo building, the units are typically all sold before he even breaks ground. Soon after the building is completed, he passes it off to the condo association, which is like a corporation that runs the building. And they don't care about it, because most condo owners keep their units for about four or five years—as long as they think they can get their money out, who cares what becomes of the place in twenty years? Home mortgages have a similar problem. Banks give out thirty-year loans, but in most cases they are quickly sold off and sliced and diced and securitized. At that point, nobody will hold a mortgage for more than a year or so before swapping it with some other bank or financial firm. Why should these firms worry about what's going to happen to a house or a condo a decade in the future?

The condo boom in Miami and other hotspots like Atlanta and Austin has been fueled in part by young people wanting to live in a sunny city. But in Miami, most of the cash is from overseas. Foreign nationals bought nearly $6 billion worth of real estate in Miami-Dade, Broward, and Palm Beach counties in 2015, more than a third of all local home spending there. Most of these people pay in cash. Cash deals accounted for more than half of all Miami-Dade home and condo sales in 2015—double the national average. The money comes from Venezuela, Brazil, Argentina, Russia, Turkey, you name it. Some of it is clean money, but some not-insignificant part of it is also dirty money, laundered through offshore accounts and other financial shenanigans. For foreign investors, a Miami condo is like a safe-deposit box—park your money here and it's protected by the US legal system and the basic trustworthiness of the American economy.

Being tied to foreign investment makes the Miami condo market different from the markets in other places (although, to be sure, other big US cities like San Francisco and New York all attract plenty of foreign capital). For one thing, the strength of the Miami market largely depends on the strength of foreign economies—especially those that are tied to commodities like oil. It's one of the great ironies that when the oil and gas barons of Russia and Brazil make money, they have been sinking it into Miami, a city that is literally drowning as a result of the combustion of the fossil fuels that made them rich. The whole point is that Miami is considered a safe investment. But what if it's not a safe investment? Foreign money can flow in quickly, as it has in Miami—but it can flow out even faster.

Add all this up, and you get a real estate game in Miami that looks something like this: In a winning scenario, civic leaders address the risk of sea-level rise in a proactive way, lobbying hard for state and federal funds and demonstrating enough political courage to raise taxes so that the city will have the money to elevate streets and causeways, invest in better sewer systems, and keep the low-lying airport functioning smoothly. Foreign investors don't panic, property values don't plummet. Population declines and some buildings are abandoned, but innovation flourishes and new ways of living with water emerge—houses float, canals replace streets, rooftops host gardens. The water keeps rising and people keep leaving, but it is a slow, stable retreat buffered by waves of innovation and civility.

In the losing scenario, the more investors understand the

risk of sea-level rise to buildings and infrastructure, the less willing they will be to invest in the region. As people sell, the supply
of houses and condos rises and prices fall. Property tax revenues decline. Even a modest drop has enormous consequences
for the city and county budgets. This means cuts in teachers and
cops and firefighters, but it also means less money to buy pumps,
fix roads, build seawalls, and build and maintain all the other
infrastructure needed to deal with rising seas. Instead of having
the courage to raise taxes to make up for the shortfall, politicians, fearful of spooking the market further, fight to keep taxes
low. With no money for repairs or upgrades, infrastructure
crumbles. And that, in turn, causes more people to sell, and the
downward spiral continues. People with money leave, pirates
and con artists arrive. Instead of innovation and civility, you get
crime and lawlessness. Long before Miami is the New Atlantis, it
will be broke and waterlogged and full of half-abandoned
neighborhoods where mosquitoes breed and leaking septic systems turn Biscayne Bay into an algae-filled lagoon.

But there's one other factor in this roulette game that
could have a big impact in Miami, especially in the inland
areas where, as one condo analyst said to me, "the people
who care about the price of diapers live."

It's called flood insurance.

In the fall of 2016, I stood on the seawall in St. Augustine,
Florida, and watched Hurricane Matthew push the Atlantic
into the historic city. Over the course of about two hours, the
storm surge overtopped the seawall and flowed across
parking lots, down streets, and over curbs, rising higher and
higher. Within a half hour, the old city was under several feet
of water. It was frightening to witness how fast it happened.
As I drove around the abandoned-feeling city, the cold, dark

water flowed everywhere, knocking down trees, swirling into and around buildings and homes, carrying branches and debris. The only people I saw were faces in upstairs windows, looking out at the rising water, and a police officer sitting in his car on the Bridge of Lions, blocking access in case anyone was so foolish as to try to get to the beach.

The next morning, the water was gone. People were out surveying the damage. The local radio station was already streaming endless commercials by sleazy lawyers offering help, as one ad put it, "to secure maximum compensation" from the government and insurance companies for storm damage.

This is, of course, an old story in Florida. The biggest risk of living on the Florida coast has always been hurricanes, as residents learned in 1926 and again in 1992, when Hurricane Andrew caused $25 billion in damage. Every year, when hurricane season starts, the question hovers out there—Is this the year? Sea-level rise only increases the risk of damage from hurricanes: the higher the ocean is to begin with, the higher the storm surge will be pushed onto land. When Hurricane Sandy hit New York City in 2012, increased wave height caused by sea-level rise resulted in about $2 billion in additional damage to the city.

Property owners deal with this risk by purchasing hurricane insurance. But hurricane insurance only covers damage from wind events, not water. If a building is damaged by surging waters that were *driven* by wind (i.e., a hurricane), then yes, it may be covered. But in most cases flooding from big rains or swollen rivers or high tides is excluded.

In the United States, virtually all flood insurance is provided through the National Flood Insurance Program, which was created in 1968 in the wake of Hurricane Betsy, which caused massive flooding in the Gulf states. In the aftermath, many commercial insurers refused to sell insurance to people who lived in flood zones. To fill the gap, and to give protection to the often poor homeowners who lived in low-lying areas, the NFIP was born.

For the NFIP, flood risks are determined by the Federal Emergency Management Agency, which draws up maps of areas likely to flood based on ground elevations and other factors. Every building within those flood zones that has a mortgage backed by Freddie Mac or Fannie Mae—that is, the two government-sponsored agencies that securitize mortgages made by banks and other lenders—must carry flood insurance. For practical purposes, this means that every building with a mortgage in a low-lying area near the coast or a river must carry flood insurance. Right now, the maximum allowable coverage under NFIP is $250,000 for residential properties and $500,000 for commercial properties.

NFIP was a good idea at the time. But it has grown into a bureaucratic, outdated, mismanaged program that subsidizes insurance rates for homeowners who live in high-risk areas. Whatever its virtues, the program has encouraged building in flood-prone areas and conditioned a generation of American homeowners into thinking that a cheap rate for flood insurance is their natural-born right as US citizens. In 2012, Congress passed bipartisan legislation to reform the program, including a number of measures that would allow insurance rates to rise to better reflect the true cost of the risk. But the political outrage was so ferocious that even the bill's two sponsors—California Democrat Maxine Waters and Illinois Republican Judy Biggert, neither of whom is known for her political cowardice—voted to repeal it within a year. Since then, Congress has tinkered with some of the rules, but the program is still woefully outdated, using flood maps that have only a vague connection to reality and do not even factor in future sea-level rise. By the time Hurricane Harvey hit Texas in 2017, causing $125 billion in damage, the program was $23 billion in debt. Congress cancelled most that debt as part of a $36.5 billion post-Harvey disaster relief bill, but the NFIP's fundamental problems remained.

"The NFIP is bankrupt," Wayne Pathman told me, sitting in his office with the Port of Miami behind him. "When I think about the future of South Florida, it's flood insurance that scares me the most." As Pathman knows as well as anyone, more than 1.7 million people in Florida have flood insurance policies—the most in America. Those policies cover roughly $428 billion in property value. Miami-Dade County alone contains 346,742 policies, which protect about $74 billion in assets. (Miami-Dade contains more policies than every state except Florida, Texas, and Louisiana.)

To illustrate the risks, Pathman showed me a few slides from a presentation he gives to investors and civic groups in South Florida. The most dramatic slides show the economic impact of a steady rise in flood insurance rates on yearly premiums and home values. In one of Pathman's examples, for a house that is worth $350,000, a homeowner might pay $2,500 a year in flood insurance. If rates go up by 18 percent a year (the maximum allowed by current law), then in ten years, the homeowner will be paying more than $11,000 for insurance. At the same time, higher premiums decrease the value of the home. If rates went up 18 percent a year, a home that was worth $350,000 in 2016 would lose $172,000 in value and only be worth $177,000 in 2026.

"That would just kill the real estate market here," Pathman said, point-blank.

Recently, Congress has allowed modest reforms in the NFIP, and insurance rates have started to go up—but nowhere near fast enough to make the program solvent. Still, rising rates are a big challenge for people who own property in flood zones. To make matters worse, the rates are dependent on some mix of luck, bad mapping, and political influence. I met one resident of South Miami who lives in a house that is ten feet above sea level and worth about $500,000. Flood insurance wasn't required when he bought it in 2002. But then FEMA decided it was in a flood zone and hit him with a bill for $600. He contested the amount, arguing that the house was higher than they thought, which knocked the bill down to $275. But in the last few years, it has gone up again, to $475. In another part of Miami, not far away, I met a middle school teacher who pays $1,873 a year for flood insurance on a house that is worth about $350,000 and is seven feet above sea level. A lawyer I know has a house at the same elevation that is worth $22 million, and he pays $600 a year. Another friend has a house eight feet above sea level in an unfashionable neighborhood south of Miami. His house is only worth $250,000, but he pays $2,500 a year for his flood insurance.

Often mayors and civic leaders argue to have neighborhoods removed from designated flood zone areas simply to keep the real estate market alive. In St. Augustine, ten thousand properties had been removed from a flood zone a few weeks before the hurricane. Many of them were inundated. In 2015, large parts of New Orleans, which are below sea level and protected only by levees, were removed from flood maps. In Broward County, Florida, which is just north of Miami and just as vulnerable, two hundred thousand people's properties were recently removed from flood zones.

NFIP is such a disaster that Congress ultimately will have no choice but to reform it, allowing rates to rise and more accurately reflect the risk. But as Pathman explained, rising rates are not the only thing that is going to change in years to come. "So far, banks don't require any more than the minimum of insurance to get a mortgage," Pathman said. "But in the next decade or so, as the risks of flooding from sea-level rise get clearer, that is probably going to change. They will begin to require that some larger percentage of an asset be insured. Banks will say, 'What you have now doesn't cover the risk. We need insurance for thirty to fifty percent of the value of the property.' If you have a two-million-dollar home, you will need to carry eight hundred thousand dollars in insurance. What happens if insurers don't want to write that? Maybe they stop giving thirty-year mortgages. And if that happens, this city is in big trouble."

As Pathman knows, private insurers are starting to get into the flood insurance market now, using sophisticated mapping technologies to pick off low-risk properties and sell their owners policies at competitive rates. But there is no question that as waters rise, insurance rates will go up and up and up. And that is likely to have a profound effect on places like Miami. As Alex Kaplan, a catastrophic risk expert at Swiss Re, the global reinsurance company, told me, in a masterpiece of understatement, "When people have to pay more and own more of the risk themselves, their decisions about where and how they live will change."

The city of Sweetwater, like most of Miami-Dade County, used to be part of the Everglades. It was a paradise of mosquitoes and alligators, entirely uninhabitable for humans until canals were built in the early twentieth century and the swamp draining began. In the late 1930s, a group of Russian-born circus dwarfs were looking for a place to call home and settled in Sweetwater. The Royal Russian Midgets, as they were called, had big plans, including the development of a Russian midget tourist attraction. It never happened. Today Sweetwater is a city of 21,000 people, mostly Hispanic, with a median household income of $32,000, making it one of the poorest cities in Miami-Dade County. It is also one of the most corrupt. The *Miami Herald* called Sweetwater "ground zero for Miami-Dade sleaze," chronicling a long history of dirty cops and city commissioners. In 2014, Mayor Manny Maroño was sentenced to three years in prison for accepting bribes.

Sweetwater is twenty miles from the Atlantic Ocean, so you wouldn't think sea-level rise would be an issue. But it is. For one thing, Sweetwater sits in a particularly low spot in the region. When it rains, it floods. And because there is no municipal sewer system in Sweetwater, the flooded water carries bacteria from septic system drainage lines, potentially creating a public health hazard. As the seas rise, the water in the drainage canals that open into the bay will rise with them, making the flooding problem in the city worse. More important, water levels in the Everglades, which are just a few miles to the west of Sweetwater, will rise too—and that means the city will flood from both sides. In the not-too-distant future, city officials in Sweetwater may be faced with the same decision that officials in other South Florida cities are confronting: elevate streets, buildings, and critical infrastructure. Or stand by as property values plummet and people leave.

I spent a lot of time in Sweetwater while I was reporting this book. I met engineers with the South Florida Water Management District who were raising the walls of the canals to help reduce flooding. I ate lunch at Nicaraguan and Cuban restaurants and talked to waiters and waitresses about flooding and sea-level rise. I talked to people who were washing their cars and walking their dogs. And my conclusion from all this is that the vast majority of the people who live in Sweetwater have no idea about the risks they face from rising waters. Most of the people I talked to were working two jobs—balancing kids, elderly parents, medical bills, car troubles. They had no time to worry about the future.

But Xavier Cortada does worry. "I'm afraid my people are going to lose everything," Cortada told me as we drove through the city one sunny afternoon. Cortada is a well-known Miami artist who spends a lot of time trying to raise awareness of the risks of sea-level rise in working-class communities that are far from the glittering coast. I had spent enough time with Cortada to know that when he said "my people," he meant not just the people in Sweetwater, but also the people in nearby Hialeah, Cuban immigrants, Brazilian immigrants, African-Americans—basically everyone who struggled to get by, living in a flimsy stucco building and driving a car with a rusty undercarriage and working to feed their families with a bank account filled with zeroes.

Cortada was then fifty-two. He is a solidly built man with a large round face and short gray hair; he is openly gay. He somehow manages to grasp the full extent of the tragedy that is facing us, while at the same time maintaining a buoyant and almost cheerful manner. He is the child of Cuban refugees and, before he became an artist in his late twenties, had been a law school student, a street gang counselor, and a mental health counselor. He thinks of art as a continuation of that work—a way to raise awareness and make people think differently about the world around them. He has painted mangroves on city overpasses, created banners that celebrate the discovery of the Higgs boson at the Large Hadron Collider in Switzerland, and led a ten-year tree-planting campaign for schoolchildren in Miami.

Given his background, it's not surprising that Cortada was acutely aware of the parallels between political refugees of the past and climate refugees of the future. In his painting called *Testamento,* words from the will of a Cuban grandfather and words from the property deed of a Cuban-American granddaughter are depicted sinking beneath the waves. The grandfather's property was lost to his grandchildren because of the rise of Fidel Castro and his Communist regime; in a similar way, Cortada believes the woman's house will be lost to her grandchildren because of the rise of the seas. "We are all displaced people here in Miami," he told me.

As we drove through Sweetwater, Cortada pointed to strip-mall pawnshops and the low stucco apartment buildings and a few little flat-roofed bungalows. "Despite the cor-

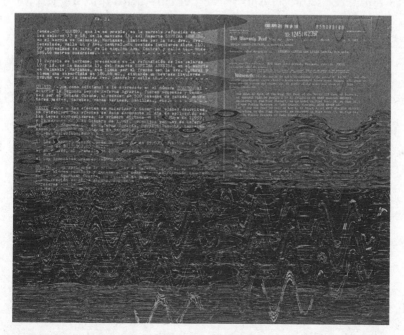

A section of *Testamento* by Xavier Cortada, depicting property deeds being lost beneath the waves. *(Photo courtesy of Xavier Cortada)*

rupt politicians, this is a place where the vast majority of people still believe in the American dream," he said. "They still believe the way you get ahead is by working hard, saving all your money, then buying a house and working to pay it off. If you get lucky and make some money, you buy a second house. These are the people who are going to get screwed. If you have all your wealth tied up in your house, and your house is underwater, then you have lost everything."

We drove onto the campus of Florida International University, which is at the edge of the city. Cortada had arranged a group discussion about sea-level rise with a small group of community members—students, the city manager, teachers. They talked about being scared whenever it rains. They talked about the city's $17-million budget, nearly all of which went to cops and schools and elder care, with nothing left for "luxuries" like improved septic systems. Janet Olivera, the principal of Sweetwater Elementary School, described her desire to instill bigger dreams in her kids, 93 percent of whom come from families with incomes below the poverty level. For her students and her school, flooding was just one problem among many. "Some of our kids had to paddle out of after-school classes in canoes," she told me.

When the meeting was over, Cortada and I walked back to his car. He seemed shaken up. We sat in the parking lot. "What happens when the city commissioners have to choose between senior lunches and flood protection?" he wondered aloud. "What happens when they try to increase the tax rate? Where is the money going to come from? What happens when real estate prices start falling? What happens when people start abandoning their homes? Who do you think is going to bail them out? Nobody. When the water comes, it will come everywhere, to cities up and down the coast. Everyone will be crying for help. Who is going to care about Sweetwater?"

There are people with money, of course, who might want to save Sweetwater when things get desperate. If the price was right and zoning restrictions loosened, they might buy up entire city blocks, knock everything down, build more expensive apartments and condos targeted at FIU students. These new buildings would be higher and more resilient and might buy a few more decades. Then again, maybe not. Maybe putting money into a low-lying place like Sweetwater will just seem foolish, and people with cash will buy up homes for pennies on the dollar and rent them out to poor people who can't afford to live anywhere else and just let the houses fall into the water, returning the land to the alligators and mosquitoes that lived here before the Russian midgets arrived.

Who knows? But when it comes to dealing with the impacts of sea-level rise, two things matter: money and elevation. Sweetwater has neither.

Eventually Cortada started up his car, and we drove along in silence for a while. The sun was going down. Afternoon traffic was heavy. "I'm worried about what the human response will be to this," he said as he drove. "If there is enough time and if we can start planning now to make a graceful retreat, it might be manageable. But if people panic and start selling their homes and leaving the city and it becomes every man and woman for themselves, it will be a disaster. My biggest fear is mayhem—the *Mad Max* response."

"How about you?" I asked a few moments later. "Are you making plans to sell your house and get out of South Florida?"

"No, not me," he said, inching his way onto a highway full of distracted commuters. "I'm never selling. I'm going down with the ship."

6. THE FERRARI ON THE SEAFLOOR

O VENICE! THE moment I stepped into the vaporetto at the airport and headed across the lagoon, I felt the romance of this ancient city. Sleek wooden water taxis speeding by like escapees from a Fellini movie, fresh sea air, church spires hovering in the blue-green distance. Of all the water cities in the world, Venice is the one that on the very first glimpse shuts down the brain and suggests what a colossal error it may have been for humans to stake their claim on the land rather than the sea.

I disembarked at the Fondamente Nove vaporetto stop and walked the short distance to my hotel, which was in an old monastery in a quiet part of the city. On my way there, I crossed over a canal, where a few boats were tied up against the buildings. They were not gondolas but working boats, the Venetian equivalent of a Ford pickup. I paused for a moment and was immediately struck by the quietness of the city—there were no people around; the shutters of the buildings were all closed. Even better, there were no cars—no traffic, no exhaust fumes, no roar of broken mufflers. The view down the canal from the little bridge looked like it hadn't changed in five hundred years. I thought about a few lines by the poet Joseph Brodsky, who had written a long essay about Venice that I had read on the plane. "Time is water and the Venetians conquered both by building a city on water, and framed time with their canals. Or tamed time. Or fenced it in. Or caged it." I'm not sure Brodsky would say the same now—but I certainly felt like I had passed into another world.

Later that afternoon, I asked the desk clerk at my hotel about the best route to Piazza San Marco. "Just follow the people," he said. And I did. As I walked through the narrow streets into the center of the city, I joined a trickle of tourists, which soon became a river and finally became a raging torrent of humans, shoulder to shoulder, bumping up against one another as they stared into shop windows or down at their iPhones. At some point I lost all sense of direction and just went with the flow—whether I would wash up in Piazza San Marco or Piazza Banana Republic I had no idea. But then the crowd broke and there I was in the famous square, the Byzantine towers and spirals of the basilica rising beside me, the stones of the piazza stretching before me like a medieval prairie.

I was surprised to see water pooling in the square. Most of Venice is only about three feet above sea level; parts of the piazza are less than two feet above sea level. Still, high tide was an hour away, and I thought the piazza would be dry at least until then (it had been a bright, clear day). I ordered a prosecco in one of the outdoor cafés and watched the pools of water expand as the tide rose. It was coming up out of the storm drains and between the stones in the piazza. Some

High tide in Venice inspires innovative transport systems. *(Photo courtesy of Anna Zemella)*

parts of the piazza remained dry; other parts were several inches deep. By the time I finished my drink, the water was nearly to my feet. Hundreds of tourists were milling around, taking selfies in front of the basilica. Across the piazza, a jazz band played. No one panicked. No one seemed too concerned about the water.

I walked over to explore the piazza. In some places, the water was ankle deep. People just detoured around it. I wandered over to the famous winged lion, which has stood high on a column here since the eleventh century. All along the edge of the lagoon, waves lapped over the edge of the quay. It looked like the lagoon was going to start waterfalling into the city at any moment. I checked the tidal app on my iPhone, which gave me real-time readouts from the city's tidal gauge— tonight's high tide was about average for this time of year, nothing unusual. Just another night in a sinking city.

Venice was founded in the fifth century, as the Roman empire crumbled and the Goths arrived to sack it. People of the Veneto, a prosperous area of rolling hills and fields, fled to the lagoon for safety and freedom. On the low, marshy islands, they built shacks, fished, and harvested salt from the lagoon. These early Venetians didn't just live on the water— they were in some sense born from the water. As Venetian humanist Giovanni da Cipelli famously put it in the sixteenth century: "The city of Venetians according to divine providence was founded on water and by water surrounded and by water protected as by a wall."

From the beginning, flooding was commonplace. The first mention of *acqua alta* dates to the eighth century, when it was recorded that "there was so much water all the islands were submerged." Early Venetians elevated their buildings on timbers; as the city gained wealth and power, the timbers increased to support their grand palazzos. Over the centuries, Venetian builders acquired a lot of wisdom about how to build in a saltwater world. For example, they knew that the timbers needed to be entirely submerged; if they were exposed to air by the daily ebb and flow of the tides, they would rot. So Venetian builders drove the timbers beneath the water and capped them with stone. Wisely, they used a particular type of stone, which comes from Istria, in what is now Croatia, which looks like marble but is in fact a dense limestone and (unlike marble) is almost impervious to salt water.

As time passed, Venetians continuously elevated the city above flood levels. In many cases, they accomplished this simply by building on top of other buildings, turning Venice into a kind of architectural layer cake. The height (or, if you will, depth) of this layer cake still surprises historians. A few years ago, during a renovation of the Malibran, one of the great opera houses in Venice, workers dug down and found that the theater had been built right on top of Marco Polo's house, which was built in the thirteenth century. They kept digging and came to the ground floor of another house, which was more than six feet below ground level today. Below that, they found an eleventh-century floor. Below that, an eighth-century floor. And below *that*, a sixth-century floor.

But Venice's centuries-old equilibrium with the sea eventually became unbalanced. Groundwater pumping by industries on the fringes of the lagoon caused the ground beneath the city to subside, exacerbating the flooding problems. Dredging the lagoon to create a channel for container ships heading into the commercial port changed the tidal dynamics, making the city more vulnerable to storm surges.

On November 4, 1966, gale-force winds in the Adriatic Sea pushed a wall of water into the lagoon. Venetians awoke to find their city under five or six feet of water. Electricity was cut, heating-oil tanks swamped, ground floors submerged. Strong winds kept the water in the city for the entire day. When it finally retreated, it left the city full of broken furniture, wet garbage, dead animals, and raw sewage. Miraculously, no one died.

"It is difficult to overestimate the effect that the flood of 1966 had on the way people still think about Venice," historian Thomas Madden wrote. "Today the most common opinion, held even by people who know nothing else about Venice, is that it is sinking. Before 1966 this opinion scarcely existed. The devastating flood had cast Venice in an entirely new light. It had always been a fragile place of exquisite beauty and slow death. It was now an emergency. Venice was descending beneath the all-consuming waves, and something needed to be done—immediately."

And it was. UNESCO, the cultural and scientific arm of the United Nations, opened an office in Venice; groups like Save Venice and Venice in Peril sprang up, raising tens of millions of dollars from around the world to save the crumbling frescoes and old church façades. More important, the Italian government passed the Special Law of 1973, which guaranteed funds to preserve "the historical, archaeological, and artistic environment of the city of Venice and its lagoon." A ban on groundwater pumping halted the city's subsidence (or at least slowed it to its current rate, one millimeter or so a year).

But it was clear that Venice needed greater protection from the sea. To come up with solutions, a group of engineering firms banded together with the government to form Consorzio Venezia Nuova (Consortium for a New Venice). Six options were proposed, from simply raising the pavements in the city to building an enormous levee to wall the lagoon off from the sea. But the idea that won out was the construction of high-tech mobile barriers at the inlets of the lagoon that would rise to protect the city when a storm approached, then lower to allow the lagoon to remain connected to the sea. In 1994, Italy's Higher Council of Public Works approved the plan. Engineers gave it the name Modulo Sperimentale Elettromeccanico, better known as MOSE, an acronym deliberately designed to invoke Moses, the Bible's great parter of the waves. Consorzio Venezia Nuova was awarded the contract to build the barriers without opening the process to bidding from other firms (a move that many Venetians were later to regret). Prime Minister Silvio Berlusconi, the brash media tycoon and center-right leader who later was convicted of tax fraud, pushed hard for the barriers and laid the first stone for the project in 2003.

The MOSE barrier is an ambitious and sophisticated piece of engineering. It's actually three separate flood barriers, one at each of the three inlets of the lagoon. Each barrier is made up of about twenty individual gates, which are bound by a hinge on the floor of the lagoon and are hollow, allowing them to fill with water. In calm weather, the water-filled gates sit at the bottom of the lagoon. But when an exceptionally high tide threatens—Venetians call it *acqua alta*—the water is pumped out of the gates and replaced with air, allowing the gates to float up to the surface and create a barrier capable of stopping a storm surge as high as ten feet. When the surge passes, the air inside the gates is released and replaced with water, causing the gates to sink back to the bottom of the lagoon.

As mobile barriers go, MOSE is sleek and elegant. Mobile barriers are to sea-level rise as condoms are to sex: a device you use to protect yourself in a heated moment. Unlike an actual wall or dike, a mobile barrier like MOSE is designed

Venice's MOSE barriers are designed to rise and fall with the tides. *(Photo courtesy of Shutterstock)*

to be deployed only when necessary to protect the city from a surge. It doesn't cut Venice off from the sea. It doesn't impede tidal flow in the lagoon. It's not a monstrous industrial structure on the Venetian horizon. And yet, if a big storm is coming, it can be raised in thirty minutes, creating a temporary wall against the water.

That's how it is supposed to work, anyway. In reality, the project has been delayed, tangled in a high-profile corruption scandal, and victim to engineering problems that have led many thoughtful observers to wonder if it will ever work. To top it all off, cost overruns have raised the price tag from $2 billion to $6 billion—and counting. At one conference I attended in Venice, Pier Vellinga, a climatologist at Wageningen University in the Netherlands, called the MOSE barrier "a Ferrari on the seafloor." It wasn't clear whether he meant this as a compliment.

As it happened, I arrived in Venice on the fiftieth anniversary of the 1966 flood. Posters advertised commemorative events throughout the city. The front page of one newspaper included pictures of people paddling in boats in front of the basilica and walking through the piazza in chest-high water.

That afternoon, I attended an event at a museum on Piazza San Marco, where the mayor of Venice, Luigi Brugnaro, a right-wing businessman ("our little Trump," one member of the audience whispered to me), addressed a small crowd of Venetian businesspeople and politicians. A high-tech video of the MOSE barrier played on a monitor behind him, with images of the gates rising and falling, highlighting the complex engineering of the caissons on the seafloor where the gates will be hinged. The video was slickly produced, as if it had been developed for the latest Apple product. Brugnaro accompanied the video with a highly animated speech. I don't speak Italian, so I guessed he was enthusiastically touting MOSE.

I leaned in to Jane Da Mosto, an activist and a longtime resident of Venice, and asked her what he was talking about. "Money, money, money," she said. "He is saying we need more money."

Afterward, I walked out on Piazza San Marco with Da Mosto. She was stylishly dressed, with a colorful scarf thrown around her neck, and had wavy brown hair and blue eyes. She is a civic activist, a scholar of sea-level rise, a mother of four, and a member (by marriage) of one of the oldest families in Venice, dating to the fourteenth century. A few years ago Da Mosto started We Are Here Venice, a citizens' group of Venetians concerned about the future of the city. She had put together a number of events to commemorate the fiftieth anniversary of the 1966 flood, including an exhibit of photographs, trying to raise awareness of what is going on now. She also persuaded shop owners in the piazza to commemorate the 1966 flood by taping a blue line on the front of their shops that marked the water level of the flood. For many shops in the piazza, the blue line was waist-high—a vivid reminder of what's at stake.

Da Mosto and I talked with a few shop owners in the piazza. Much of the discussion was about another kind of flood—the flood of tourists, especially those brought in by the giant cruise ships that now invade the Venice lagoon like beasts from another planet. About 20 million tourists visit Venice every year, overwhelming the historic city's 56,000 residents. "We are drowning in tourists," one shopkeeper told me. "But we need them to survive."

As Da Mosto and I walked around the piazza, she told me that the cruise ships and sea-level rise are the two most power-ful threats that Venice faces right now. The cruise ships have transformed the Venice economy into a singular engine that services tourists: every shop sells necklaces of fake Murano glass jewelry and Venetian carnival masks; every restaurant offers the same pasta with meatballs; every apartment is now an Airbnb. This has not only eroded the city's tax base and pushed out traditional jobs, it has turned the city into some-thing nearly indistinguishable from a Disney version of itself.

It also distracts from the larger threat of sea-level rise. As Da Mosto pointed out, the flooding is getting worse and worse. In the 1940s, Venice flooded only about ten times a year; now it floods seventy-five times a year. "Of course, Vene-tians are so used to living with water, they hardly notice the change," Da Mosto explained. "Many people just think this is how it has always been."

And they are partly right. Maintaining Venice has been a constant challenge ever since the city was built, but the onus is only getting heavier. As the water rises, it floods above the Istrian stone at the base of many of the historic buildings and walls. The salt water seeps into cracks and is sucked up into the brick and marble, both of which are highly vulnerable to corrosion. As the walls and foundations lose their structural integrity, windows and door frames sag, ceiling beams weaken. Various chemical and physical treatments have been developed to protect the brickwork and block salt invasion. One example: the use of outer plasterwork as a "sacrificial layer" in which the salts concentrate and crystallize, rather than staying within the brickwork. This helps, but it's hardly a long-term solution to saltwater inundation.

Salt water is eroding plaster and brick in Venice. *(Photo courtesy of the author)*

Da Mosto explained that the situation is especially delicate for St. Mark's Basilica, the eleventh-century Byzantine church in the piazza that is one of the most iconic buildings in Venice. It sits at a low spot in the piazza, and the inside of the church is frequently inundated, especially the atrium. Worse, researchers have found evidence of saltwater corrosion more than twenty feet up the walls of the basilica, which has caused tiles in the centuries-old mosaics on the walls and ceiling to loosen and fall.

What to do? The floor of the basilica has been raised many times in the past, and the entire piazza has been repaved. But today this is no longer an option: further raising would destroy the architectural and compositional relationship between the buildings and between the buildings and the paving itself. "The MOSE barrier," Da Mosto said, her voice dripping with skepticism, "is supposed to be the fix for all this."

Da Mosto didn't have a lot of faith in MOSE. She believed the project was poorly engineered, corrupt, and ridiculously expensive both to build and operate. She also believed that, thanks to cost overruns and corruption, there was a good chance it would never be completed. "But what do I know?" she asked before she said goodbye and headed to another meeting. "I am not an engineer. I am just some woman who loves Venice and cares about its future."

Later that evening, as the tide rose and Piazza San Marco started to fill up with water again, I went to a special mass in the basilica commemorating the 1966 flood. In the atrium, the marble is corroded by salt water and looks like a wedding cake left out in the rain. The mosaic floor undulates; it has been flooded so many times that it seems to have permanently acquired the form of the sea. Nevertheless, this basilica with its magnificent gold dome has been here for a thousand years. It's easy to see why many people believe it will be here for another thousand.

When the mass concluded, I walked over to La Fenice, the famous Venetian opera house that had burned to the ground in 1996 and whose resurrection seven years later was cited by many Venetians I talked to as evidence of the city's resilience and ability to overcome the worst tragedies. An opera, *Aquagranda*, was created especially for the fiftieth anniversary of the flood, and did its best to re-create, in what might be called a symbolic opera documentary, the experience of living through the 1966 inundation. I'm not sure how successful it was as an opera, but it was artfully presented, with singers in rubber boots splashing through water on the stage or trapped behind a wall of water that flowed between two large pieces of Plexiglas that dangled over the stage. A burly baritone sang as the storm (and the music) gained momentum, the supertitles capturing the terror that Venetians faced that day—and may again in the future:

> *I'm scared of it*
> *No, I'm not scared of it.*
> *The water*
> *The flood*
> *The angry God*
> *Black water*
> *Vile water*
> *Dirty water*
> *Damned water*
> *Cold greedy*
> *Dirty filthy*
> *Flood*

The following morning, I visited the offices of Consorzio Venezia Nuova, the engineering group responsible for building MOSE. The offices are in the Venice Arsenal, a restored medieval boatyard with beautiful arched colonnades. Founded in the twelfth century, the arsenal mass-produced warships during the glory days of the Venetian republic and was the largest industrial complex in Europe before the Industrial Revolution (Galileo worked as a consultant here, helping shipbuilders with engineering problems). It is somehow fitting that the MOSE project is housed at the arsenal. After all, the MOSE barrier is to Venetians today more or less what the galley ship was in the fifteenth century: a tool to protect the city from invasion.

The CVN offices, in a restored brick boatworks building, have glass-walled cubicles and sleek conference tables. They're very Silicon Valley–esque. There are posters of the MOSE barriers on the walls, and the quiet is broken by random beeps of email popping in on someone's computer down the hall. I met with Monica Ambrosini, one of CVN's media handlers, in her office. Over an espresso, she explained the current state of construction: so far, gates had been installed in only one of the three lagoon inlets. I asked her if she could take me out to see the installation, but she shook her head. "It has been tested, but it's not operational now, so there is nothing really to see. It is all underwater now." She explained that she would like to take me to see the control room of the MOSE barrier, but unfortunately it was not built yet. "But I can show you a couple of gates that have just arrived," she offered.

Ambrosini was eager to show off progress on MOSE in part because the organization has taken a lot of fire in recent years for wild cost overruns, as well as a corruption scandal that left many Venetians wondering how much money went into the design and construction of the barrier and how much went into the design and construction of vacation homes for the politicians who were involved with the project. In July 2013, five hundred police officers raided a hundred and forty offices from Venice to Tuscany and Rome as part of a probe into the alleged rigging of contracts. More than thirty-five people were arrested, including politicians and business execs who were accused of bribery and other forms of corruption. The investigation eventually led to the indictment of the mayor of Venice, as well as the governor of the Veneto region. As one Italian economist argued, MOSE was "more beneficial to those who were awarded monopolies to build it, and to politicians who used it for illicit gains, than to the citizens for whom it was supposedly designed." No one knows for sure how much was siphoned off by corrupt officials, but a good estimate is close to a billion dollars.

When I asked Ambrosini about all this, she blushed and asked if she could go off the record—then explained, in so many words, that she couldn't justify or explain it but that bad people had been involved and now those bad people were gone. Back on the record, she said, "We have completely changed the structure of the organization and put new people in charge and new rules in place so it cannot happen again."

With that, we put on our hard hats and orange safety vests and head out to the backwoods of the arsenal. We walked by abandoned brick buildings with weeds growing around them, and then to a large open area near the lagoon. Workers scurried around; a forklift carrying concrete pipes rolled by. Ambrosini led me toward a chain link fence, behind which were two enormous slabs of metal. They looked like hundred-foot-long, fifteen-foot-wide teeth extracted from a creature on planet Gargantua. One was painted bright yellow, the other aqua. Ambrosini pointed to the aqua one. "It weighs about three hundred thirty tons," she explained. The mobile barrier at the central inlet would be made up of twenty-one of these gates, each hinged to the ocean floor and working together.

That was when the scale and ambition of this project became clear to me. These gates were designed to do nothing less than hold back the sea, one of the primal forces of nature. That humans could even contemplate building such a tool was evidence either of the power of technological innovation or of the folly of human hubris—or both.

I asked Ambrosini when she estimated that the entire MOSE barrier would be installed and ready to defend the city.

"By 2018," she said. Then she added, "That's the hope, anyway."

The MOSE barrier may or may not be completed by 2018, but even assuming that the $6-billion project works as planned (a big assumption), the design and construction of the barrier bring up a number of issues that are worth considering.

The first is the time required to design and build the barrier. After the 1966 flood, it took more than fifty years to settle on a plan to protect the city, then get it approved, funded, designed, and partially built. That kind of time scale might not be important if you are building, say, a new city hall. But when you're constructing something that has to adapt to climate change, fifty years is like fifty centuries. In 1966, sea-level rise was something that only a few scientists thought about: today, it's an existential threat to cities around the world. But even though the design for MOSE was not finalized until 2000, the estimates for sea-level rise were still woefully out of date. According to a UNESCO report, during the project planning phase, three sea-level rise scenarios for 2100 were considered. The estimate cited as "most probable" was 16 centimeters (about 6 inches); the one cited as most "prudent" was 22 centimeters (about 8 inches); the third scenario, labeled "pessimistic," was 31.4 centimeters (about a foot). Planners recommended using the prudent scenario for design purposes. In a world where respected scientists are now suggesting that sea-level rise by 2100 could be six feet or more, designing a barrier for eight inches of sea-level rise looks, in retrospect, either like startling naïveté or startling incompetence.

The second issue is cost. When you spend $6 billion in tax dollars to fix a problem, it had better work, because at that price, it's unlikely you'll get a do-over. But costs don't stop there. There is also the question of maintenance. MOSE might be a Ferrari on the seafloor, but Ferraris require a lot of work to keep them rolling. Initial estimates for maintenance charges were $5 million to $9 million a year. Ambrosini suggested that $50 million a year is a more appropriate figure; others have suggested it could go as high as $80 million a year, depending on how often the gates are used. This money is like a ransom that has to be paid every year—if maintenance lags and the gates fail, Venice could be inundated.

The most important question, of course, is whether the barrier will actually protect the city—and if so, for how long. That is not a simple question to answer, because it depends on how you define "protect." If the barrier is completed and functions as advertised, it will likely spare the city from storm surges like the one in 1966—for the next few decades, anyway. But because the gates will only rise when the tide in the lagoon reaches 110 centimeters (about three and a half feet), they will not stop the flooding that already happens in low spots in Venice when the tide gets above 80 centimeters or so. This could be solved by raising the barrier at lower tides, but that would have a big impact on the health of the lagoon, and would also increase operational wear and tear on the barrier, which would raise maintenance costs.

In the longer run, questions about the protection of the barrier depend, like much else in this world, on how far and how fast sea levels rise during the fifty-year design-life of the barrier. Although MOSE was designed to protect from tidal surges as high as nine feet, it was only engineered to handle less than one foot of sea-level rise. This is an important issue, not only because rising seas increase the chances of higher storm surges, but also because higher seas would require the barrier to operate more frequently, as well as putting additional stresses on the barrier that it was not designed to handle.

Right now, engineers expect the MOSE barrier to be closed about ten times a year, usually for about five hours, until the tidal surge passes. Georg Umgiesser, an oceanographer at the Institute of Marine Sciences of the Italian National Research Council, estimates that with 50 centimeters (nearly 2 feet) of sea-level rise, it will be closed once a day. With 70 centimeters (a little more than 2 feet), Umgiesser's research suggests, the gates will be closed more often than they are open. More frequent closing means not just higher maintenance costs, but ever-greater dependence on the barrier to prevent massive flooding in the city. An outright failure could be catastrophic. And of course, if the gates are closed all the time, the whole point of a $6-billion Ferrari on the seafloor is moot. Why not just build a solid wall at a fraction of the cost and be done with it?

After our visit to the gates, Ambrosini and I returned to her office, where she spread out a map of the Venice lagoon. I brought up the problem of protection from sea-level rise.

"In the end, I guess the big question is, how much sea-level rise can MOSE protect the city from?"

"We believe Venice will be protected up to sixty centimeters of sea-level rise," she told me bluntly.

This was more than the MOSE design parameters suggested, but I let that go. It was still shocking to hear her admit this so bluntly. Some scientists believe we could have sixty centimeters (about two feet) of sea-level rise as early as 2050.

"And after that?"

Ambrosini's professional demeanor shifted and she suddenly looked worried. She pointed to the map, sweeping a finger over the northern edge of the lagoon. "After that, the sea will come in from other places to the north and the south. There is nothing we can do to stop it."

While I was in Venice, I couldn't help but draw parallels with a trip I had taken a few months earlier to Rotterdam for a conference on sea-level rise. Rotterdam is often cited as one of the cities in the world best adapted to sea-level rise. That is probably true. But it is also true that Rotterdam is an almost entirely new city. It was heavily bombed during World War II, and one of the most striking things to me about the city is the lack of any ancient architecture, any sense of the kind of history that one associates with an old northern European city. It is all modernist towers, square and blocky. It is also a modern architect's playground: the train station looked like a cresting wave, buildings were made of cubes and trapezoids, and I bought coffee in an enormous new marketplace that looked like an airplane hangar with flowers and cows painted on it.

Rotterdam is the largest port in Europe, strategically located on the Rhine about thirty miles from the North Sea. The city is full of old canals and wooden boats that have been transformed into hip hotels and restaurants. Flooding is a well-known threat here: the city sits on a plain that absorbs outflows from the Scheldt, Meuse, and Rhine Rivers, which makes it vulnerable to inundation from extreme rainfalls, as well as sea-level rise and storm surges that move up the Rhine.

Dutch engineers have come up with a lot of innovative ways to deal with flooding in Rotterdam, including pioneering the use of "water parks," which are public squares that double as catch basins for water, essentially creating storage ponds that keep the water from draining into the streets and flooding neighborhoods. During big rainstorms, the catch basins funnel the water into the storm water discharge system, sending it out into the river. I visited one of those water parks in Rotterdam. It was a sunny day, so the water park was dry—it just looked like a large sunken concrete plaza between modern office and apartment towers. A sequence of concrete benches led down into the catch basin in the center of the square. As far as public spaces go, it was pretty bleak. But perhaps less bleak than a flooded city.

While I watched Piazza San Marco flood in Venice, I thought about those water parks in Rotterdam (I'd also visited one in Copenhagen, which was notably less bleak, with rain-catching sculptures and water-friendly grasses and trees). If you don't care about history, or the architectural integrity of Piazza San Marco, it's easy to imagine how it could be transformed into a water park. It could be far more beautiful than the featureless concrete square in Rotterdam, a sunken piazza that would drain water from the square while still providing a place for people to gather, just as the square does now. This would save the basilica from the corrosive effects of salt water, stabilize the bell tower, and, most important, suggest that Venice is not a museum, but a living, breathing city that is changing over time.

Venice and Rotterdam do have at least one thing in common: their understanding of the risks of living with water has been shaped by tragedy. In 1953, a great storm swept in from the North Sea during high tide and flooded large parts of the Netherlands, the UK, and Belgium, killing two thousand people. In the Netherlands, a twenty-foot surge broke through dikes, flooding areas that the Dutch had believed would never be flooded. In terms of psychological impact, the catastrophe was similar to the 1966 flood in Venice. As Henk Ovink, the Dutch special envoy for international water affairs, told me while I was in Rotterdam, "It was the moment we realized we weren't so safe anymore."

Like the people of Venice, people in the Netherlands have been living with water for a thousand years. Some of the oldest laws in the country are about controlling and sharing and protecting themselves from water. Reclaiming land from the sea is, in some ways, the origin tale of the Netherlands. Thirty percent of the country lies below sea level, and this nation had long felt protected by walls and dikes against the attacking sea. But the 1953 flood was a wake-up call to the risks they faced in the future. The Dutch government responded with an all-out engineering effort to defend the nation from the sea, launching the Delta Works project, which, among other things, required all infrastructure be built to 1-in-10,000-year flood standards. That meant raising dikes and levees, rerouting rivers, and, in some cases, moving villages out of harm's way.

The centerpiece of the Delta Works is the Maeslant Barrier, about fifteen miles from the center of Rotterdam, where the Rhine enters the sea ("It's the storm drain of Europe," one engineer told me). Like the MOSE barrier in Venice, it is the central protection for the city. The Maeslant Barrier is designed to protect Rotterdam from a big storm surge coming up the Rhine and inundating the city, while at the same time keeping the Rhine open for shipping, which is important economically for all of Europe. When a storm is coming, two enormous gates swivel out from hinges on the riverbanks and close off the river from the ocean, holding back both the flow of the Rhine and the surge of the incoming ocean. The gates, held firm by 52,000 tons of concrete buried in the ground, are designed to roll with the waves, swinging on both horizontal and vertical axes like a human shoulder. However, despite the size of the barrier, it can only stay closed for about twelve hours before the stress on the joints and the foundation becomes too much and the gates have to open.

While I was in Rotterdam, I drove out to the barrier with Richard Jorissen, managing director of the Dutch Flood Protection Program. We parked near the visitors' center and walked out to have a look at the wide and powerful Rhine. We watched an oceangoing freighter pass out into the North Sea. It was carrying coal.

Unlike the MOSE barrier, which is designed to hide beneath the water until it is needed, then rise like a superhero to stop the surge, the Maeslant Barrier looms on the shore, showing off its huge hinges and enormous steel gates ("When they are closed, the gates together are longer than the Eiffel Tower is tall," Jorissen told me). Instead of hiding discreetly away, it shows off its muscle and dares the seas to rise and test it. The barrier, which was completed in 1997, cost $450 million to build. In the twenty years it's been in operation, it has only been used once (although it is tested every year).

"That's a lot of money to spend on something you use only once in ten years," I said to Jorissen.

"We're Dutch — we take protection seriously," he joked.

"So as long as this barrier works, Rotterdam is safe from flooding?"

"Well, no. There are scenarios we can imagine that would still be catastrophic."

"Like what?"

Jorissen described his nightmare scenario: A hurricane-level storm blows in from the northeast, dropping massive rainfall on Rotterdam. At the same time, the storm drives a wall of water up the Rhine — enough to flood Rotterdam, even without the rainfall. So operators of the barrier face a kind of devil's choice: open the gates to let the rain-swollen river out, or close the gates and risk letting the storm surge in. Either way, the city floods.

"Luckily, we haven't faced that yet, and perhaps never will," Jorissen told me.

Despite differences in their designs and price tags, the Maeslant Barrier has a lot in common with the MOSE barrier in Venice: both serve the complex purpose of keeping an important city safe from the sea, while at the same time keeping the city open to the sea. It is a difficult trick, and one that, in a world of quickly rising seas, probably won't work for long. With three feet of sea-level rise, the Maeslant will be no more helpful than the MOSE. Unless walls and levees are raised for many miles around it, the sea will come in from other directions, making the opening and closing of the gate moot. As one engineer put it to me, "These structures are not about solving the problem—they are about buying time until we better figure out how big the problem really is."

Another example is the barrier on the Thames just east of London. The same 1953 storm that devastated the Netherlands also sent a surge up the Thames into London, inspiring the British to build a barrier to prevent it from happening again. The barrier became operational in 1982, one of the first examples of a big retractable gate to be built. It has a *Star Wars* look to it, with rolling gates that open and close as the tides rise. And thanks to rising seas and increasing storm surges, it is being put to use more and more often: it was closed only four times in the 1980s; as of this writing, it has been closed seventy-five times since 2000. But UK officials are putting off designing a new barrier for the moment, choosing other adaptive measures until scientists have a better idea how fast the seas will be rising in the coming decades. Planners in the UK understand the essential problem with big infrastructure: it's very expensive, it takes a long time to build, and it's not very adaptable to changing conditions.

Another solution, of course, is to simply decide that holding back the seas is a risky and ultimately futile idea. As Ivan Haigh, an oceanographer at the University of Southampton in the UK, put it to me, "We are learning that instead of fighting the water, sometimes it is better to figure out ways to live with it."

Unlike, say, Lagos or Jakarta, where millions of people will be displaced by rising waters, it's hard to argue that the drowning of Venice will be a human tragedy. This is not to say that the displacement of 56,000 people is trivial, or that it won't involve real human suffering. It will.

But when you think about the loss of Venice, it's not the Venetians who are topmost on most people's minds. It's the loss of a beautiful and historic city that has played an enormous role in the development of Western civilization. It's the loss of the stones in the narrow streets where Titian and Giorgione walked. It's the loss of eleventh-century mosaics in the basilica, and the unburied home of Marco Polo, and palazzos along the Grand Canal that are so expertly designed and built that they have stood for five hundred years in the most precarious of locations. The loss of Venice is about the loss of a part of ourselves that reaches back in time and binds us together as civilized people.

The British poet Lord Byron, who lived in Venice for a few years in the early nineteenth century and once swam the entire four-and-a-half-mile length of the Grand Canal (after having sex twice that morning, he boasted), understood all this two hundred years ago:

> *O Venice! Venice! When thy marble walls*
> *Are level with the water, there shall be*
> *A cry of nations o'er thy sunken halls,*
> *A loud lament along the sweeping sea!*
> *If I, a northern wanderer, weep for thee,*
> *What should thy sons do?—anything but weep?*

Before I left Venice, I went to see Pierpaolo Campostrini, the director of CORILA, a quasigovernmental agency that oversees research by Italian universities and scientific groups that are working on the Venice lagoon. If there is a czar of the lagoon, it's Campostrini. He is an electrical engineer and physicist by training but is also deeply involved in the restoration and preservation of the city, including the basilica.

His office is in Palazzo Franchetti, near the Ponte dell'Accademia, on the Grand Canal. It has a businesslike feel, with maps and books and papers piled everywhere.

Campostrini and I talked late on a Friday afternoon, and he looked like he'd had a long week. His dark blue suit was a little rumpled, his manner a little hurried. Still, he was warm and friendly, and was clearly deeply concerned about Venice. "I was born here, and my kids were born here, and I hope their kids are born here," he told me.

When I asked him about the long-term future of the city, Campostrini started out by talking about the past. He pointed out that the Venice lagoon was created by sediments carried by rivers down from the Alps. "In their natural state, lagoons are short-lived," he said. "The destiny of the Venice lagoon was to be filled. The only reason it wasn't was because, at a specific moment in time, people decided to build a city in the lagoon. Venice decided to chase its own destiny. Rivers were diverted, and the filling of the lagoon with sediment stopped. The lagoon is now a human artifact, not really a natural system at all, although we still think of it that way."

In the past, Campostrini pointed out, Venice had no trouble dealing with rising seas. "We just kept building the city higher. This palazzo we're in today was built in the fifteenth century, but there's a thirteenth-century palazzo beneath this. And beneath that, who knows? They were not sentimental about the past. They did not worry about preserving old buildings. They just built new ones on top of old ones. And the city kept rising. But of course, we can't do that anymore. Now, there are cultural constraints. We don't want to lose the beautiful Renaissance architecture we have here. Knocking it down and building on top of it is not an option. We have to find another way to save it."

Campostrini agreed with many others whom I'd talked to that in the near term, Piazza San Marco poses the biggest problem. He grabbed a roll of paper on a table near his desk and unrolled it in front of me. It was a contour map of the piazza. Every curve marked a one-centimeter change in the surface height of the piazza. On this map, you could see what a wavy and uneven surface it was.

He pointed to the front of the basilica. "This is the lowest point right here—the elevation is only about seventy centimeters." The high point along the quay and other spots was about 110 centimeters. The 40-centimeter difference—about a foot and a half—was enough to send whatever water that got into the piazza toward the basilica, flooding it and causing corrosion and decay in the marble and brick.

"This is a real problem," Campostrini said gravely. "What do we do? We cannot raise the basilica, that is impossible."

He described a plan to upgrade the drainage on the basilica, so that when it floods, the water can run off quickly. He described another plan to pull up the ancient stones in the piazza and place a clay barrier beneath it to stop water from coming up from below (unlike the palazzos along the Grand Canal, the piazza is built on a natural sandy island that was part of the lagoon). "It will be disruptive, but it needs to be done," he said. It would also cost about $60 million.

"All this will be helpful for the short term," I argued, "but by the end of the century, Venice could be facing four, five, six feet of sea-level rise. Better drainage in the piazza will not save the city from that. Neither will the MOSE barrier, as I understand it."

Campostrini took in a big breath, then blew it out. He looked grave. "No, it will not."

"What does your vision of Venice with six feet of sea-level rise look like?"

"Well, we will have to take more drastic measures."

"Like what?" I asked.

"One thing that could be done is to build a levee all the way around the city, isolating it from the sea. It would of course kill the lagoon and cause many other problems, but it is possible to do."

"Turn it into a walled city."

"Yes. We could still have a lagoon, but it would be a fresh-water lagoon." He looked like he had just received a fatal diagnosis. "For this to work, we would have to change the septic system in the city, so there is no pollution. We would have to move the port. The city would have a different character. But it is possible."

He stared at the contour map of the piazza.

"Another possibility," he said hesitatingly, as if he were embarrassed by what he was about to say, "is to inject seawater into a clay layer about six hundred meters beneath the city and lift it up. In ten years, it may be possible to raise the city about thirty centimeters that way. I know it sounds extraordinary, but in theory, it is possible. The oil industry has experience with this kind of technology. And it works in computer models."

"But in real life—"

"Yes, in real life, there could be problems with this idea." He shook his head, like he knew it was an outlandish idea. "The city could lift unevenly. Which would not be good for buildings, I am sure."

I imagined standing on the quay in front of the Doge's Palace and feeling the city rising beneath me, tilting a little to the left, a little to the right, watching cracks appear in the palace walls.

"I'm skeptical," I said.

"As you should be," he sighed. He looked out his window, watching the crowd of tourists crossing the Ponte dell'Accademia. "I don't know how we will save Venice. But Venice has been here more than a thousand years. In the end, we are not so fragile. Somehow, we will figure out a way to manage it."

7. WALLED CITIES

ON A BRIGHT spring day, I walked along the seawall on the Lower East Side of Manhattan with Dan Zarrilli, the head of New York City's Office of Resilience and Recovery—basically, he was Mayor Bill de Blasio's point man for preparing New York City for the coming decades of storms and sea-level rise. Zarrilli, who is in his early forties, was dressed in his usual City Hall attire: white shirt and tie, polished black shoes. He had short-cropped gray hair, dark eyes, and an edgy I've-got-a-job-to-do manner. Zarrilli may be the only person in the world who holds in his head the full catastrophe of what rising seas and increasingly violent storms mean to one of the greatest cities in America. Not surprisingly, instead of musing about the beautiful weather, he pointed to the East River, where the water was innocently bouncing off the seawall about six feet below us. "During Sandy," he said darkly, "the storm surge was eleven feet high here."

As Zarrilli knows better than anyone, Hurricane Sandy,

which hit New York City in October 2012, flooding more than 88,000 buildings in the city, killing 44 people, and causing over $19 billion in damages and lost economic activity, was a transformative event. It did not just reveal how vulnerable a rich, modern city like New York is to a powerful storm, but it also gave a preview of what the city may face in the coming century. "The problem for New York is the same as it is for every coastal city," Chris Ward, the former executive director of the Port Authority of New York and New Jersey, the agency that runs New York's airports, tunnels, and other transportation infrastructure, told me. "Climate science is getting better and better, and storm intensity and sea-level rise projections are getting more and more alarming. It fundamentally calls into question New York's existence. The water is coming, and the long-term implications are gigantic."

Zarrilli turned away from the water and we walked toward the downtrodden park that separates the river from the FDR Drive, and, beyond that, the Lower East Side. "One of our goals is not just to protect the city, but to improve it," Zarrilli explained. In 2019, the city planned to break ground on what's called the East Side Coastal Resiliency Project, a ten-foot-high steel-and-concrete-reinforced berm that will run about two miles from East Twenty-Fifth Street down to the Manhattan Bridge. The project, which is budgeted at $760 million but will surely cost far more before it's completed, is the first part of a larger barrier system, known informally as the Big U, that someday may loop around the bottom of Lower Manhattan. Unlike the MOSE barrier in Venice, if and when the Big U is ever completed, it will be a solid wall—a modern rampart against the attacking ocean. There are plans in the works to build other walls and barriers in the Rockaways and on Staten Island, as well as across the river in

Hoboken. But the Big U in Lower Manhattan is the headliner, not just because it will cost billions to construct (rough estimates start at $3 billion and rise fast), but also because Lower Manhattan is the most valuable chunk of real estate on the planet, as well as the economic engine for the entire region— if it can't be protected, then New York City is in deep trouble.

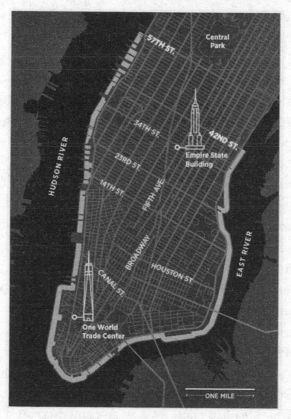

The Big U in Manhattan, from E. 42nd St. around to W. 57th St. *(Map courtesy of John Grimwade)*

Zarrilli, who doesn't like the phrase "Big U" because it sounds like a plug for BIG, the Danish architectural firm that helped design the barrier, was uneasy talking about walls, in part because that obscures other, more democratic measures the city is taking to become more resilient, such as requiring developers to elevate critical infrastructure and install robust backup power generation, but also because wall-building is politically fraught: You can't wall off the city's entire 520-mile coastline, so how do you decide who gets to live behind the wall and who doesn't? "You have to start somewhere," Zarrilli explained, "so you begin in the places where you get the maximum benefit for the most people."

In Zarrilli's view, there is no time to waste. He knows as well as anyone that even the most indomitable city in America is facing a brutal future of rising seas and increasingly violent storms. As we crossed the FDR Drive on a pedestrian overpass, I asked Zarrilli, who is the father of two young kids, if it scared him to think about the economic and political chaos that may be coming. "It's not a pretty picture, but you can't let yourself be paralyzed by fear," he said, putting a brave face on it. "You have to take it one step at a time and do what you can right now."

When it comes to sea-level rise, few cities have more at stake than New York. In purely economic terms, the New York metropolitan area is responsible for nearly 10 percent of the US gross domestic product and is the financial hub of the free world. The city also has a symbolic value that is hard to quantify, with 8.5 million people from all over the world living there, and billions more who are connected to the city by work or family or by their dreams to come here and make it big. "To deal with climate change, we need inspiration," said Henk Ovink, the Dutch special envoy for international water affairs, who was deeply involved in rebuilding New York after Sandy. "New York City is the capital of the developed world. If it does things right, it can radiate inspiration to everyone."

In a world of rapidly rising seas, New York is better prepared than many coastal cities. As anyone who has seen the rock outcroppings in Central Park knows, much of Manhattan is built on five-hundred-million-year-old schist, which is impervious to salt water. There is plenty of high ground, not just in Upper Manhattan's Washington Heights, but also along a ridge that runs diagonally through Queens and Brooklyn, including places like Jackson Heights and Park Slope. Finally, the city has brains and money and attitude—New York is not going to go down without a fight.

But in other ways, New York is surprisingly at risk. First, it's on an estuary. The Hudson River, which runs along the west side of the city, needs an exit. So unlike with a harbor city like, say, Tokyo, or a city on a lagoon like Venice, you can't just wall New York off from the rising ocean. Second, there are a lot of low areas, including the Brooklyn and Queens waterfronts and Lower Manhattan, which have been enlarged by landfill over the years (if you compare the map of damage from Sandy in 2012 with a map of Manhattan in 1650, you'll see that they match pretty well—almost all the flooding occurred in landfill areas). The amount of real estate at risk in New York is mind-boggling: 72,000 buildings worth over $129 billion stand in flood zones today, with thousands more buildings at risk with each foot of sea-level rise. In addition, New York has a lot of industrial waterfront, where toxic materials and poor communities live in close proximity, as well as a huge amount of underground infrastructure—subways, tunnels, electrical systems. Finally, New York is a sea-level-rise hotspot. Because of changes in ocean dynamics, as well as the fact that the ground beneath the city is sinking as the continent recovers from the last ice age, seas are now rising about 50 percent faster in the New York area than the global average.

*   *   *

Building fortifications around a city is an idea that is as old as cities themselves. In the Middle Ages, walls were built to keep out invading armies. Now they are built to keep out Mother Nature (or, in Trumpland, illegal immigrants). Obviously, if they are built right, they work. Seventy percent of the Netherlands is below sea level; without walls, dikes, and levees, the nation would be a kingdom of fish. New Orleans exists today only because of enormous levees holding back the sea. Japan is practically encircled by giant seawalls to protect residents from tsunamis. But even among the Dutch, the masters of Old World–style levee-building, walls and dikes and levees are falling out of favor. "We are beginning to realize we can't keep building walls forever," said Richard Jorissen, the Dutch expert who took me out to see the Maeslant Barrier near Rotterdam. "Sometimes they are necessary, but we also understand that sometimes we have to learn to live with the water. If it is not built right, a wall can create as many problems as it solves."

As far as walls go, the Big U was designed to be a nice one. It's the love child of a collaboration headed by the Bjarke Ingels Group (BIG), the Danish firm that has designed a number of playful, slightly surreal buildings around the world (including a pair of condos in Miami for developer David Martin).

The Big U was one of four winning proposals in the $930-million Rebuild by Design competition, which was sponsored by the US Department of Housing and Urban Development in the aftermath of Sandy and attracted proposals by top architects and urban planners from around the world. In an animated video that BIG created to promote the project, the Big U is depicted as an undulating public space where a grass berm is planted with flowers and trees and creates parklike areas where people can play baseball and stroll on a sunny day. The gritty, thundering empty space beneath the elevated FDR Drive is transformed into a place where kids play Ping-Pong and farmers' markets appear on weekends. The city is protected from the water by the berm (which is underlaid with steel and concrete) and walls covered in art that flop down from the FDR. It is all very cheerful and inspiring—disaster-proofing as a public amenity.

The problem is, the actual barrier may or may not resemble the barrier in the video. Several urban planners I talked to believe that, due to cost-cutting and engineering complexities, by the time it is built, the wall will be stripped of its crowd-pleasing amenities. "When it's done, it's just going to be a big dumb wall," said one landscape architect who has watched the project closely.

But dumb or not, given the amount of valuable real estate in Lower Manhattan, some kind of defensive structure is going to be erected there to keep the water out. Building a wall is (compared to more long-term and nuanced options) cheap, quick, and irresistible to politicians wanting to prove they have acted boldly. But that doesn't mean it's always the smartest or the safest solution.

For one thing, as the flawed assumptions behind the design of the MOSE barrier in Venice have revealed, there's always a question about what level of protection the barrier is designed to provide. Residents of Kamaishi, Japan, thought they were safe behind a mile-long, twenty-foot-high steel-and-concrete seawall. But when a thirty-foot-high tsunami hit the region in 2011, the seawall crumbled and 935 Kamaishi residents died. Lower Manhattan and Japan are not directly comparable, if for no other reason than the fact that Lower Manhattan is not exposed to tsunamis. But whenever you build a wall, there is always the risk that Mother Nature won't respect the design specifications. A barrier like the Big U would in theory be designed to protect from another Sandy, but not a lot more. (And by 2100, Sandy-like events are predicted to occur much more often.) I asked Kai-Uwe Bergmann, a partner at BIG, why the barrier wasn't designed to withstand a Sandy-level flood plus, say, an additional five feet to accommodate sea-level rise in the future: "Because the cost goes up exponentially," he replied bluntly and honestly.

Another obvious problem is that walls only protect the people who are behind them. The new barrier on Lower East Side will have the virtue of protecting several large public housing developments, as well as a key Con Edison substation that flooded during Sandy, causing a massive blackout in Lower Manhattan. But that barrier is likely to be just the beginning of the walling-off of Lower Manhattan. "The real purpose of the Big U is to protect Wall Street," said Klaus Jacob, a disaster expert at Columbia University. Given the importance of Wall Street to the US economy, that was not surprising. But how long do you think it will be before Red Hook, a largely poor, African-American area in Brooklyn that was also heavily damaged by Sandy, gets a barrier designed by Bjarke Ingels?

Across the Hudson River from New York City in Hoboken, New Jersey, walls posed a different problem. Much of Hoboken was built on former wetlands; when Sandy hit, water poured in like it was filling a giant bowl (one of the most iconic images of Sandy was of a sailboat beached in front of luxury apartments in Hoboken, across which someone had spray-painted GLOBAL WARMING IS REAL). To protect the city, Mayor Dawn Zimmer supported building a Big U–like barrier along the waterfront. The problem is, to protect the city, the wall has to run in front of luxury lofts with prized views of Manhattan. "I'd rather flood than stare at a wall," one Wall

Street analyst who lives along the waterfront told me. Zimmer, who, during a recent walk through the city, was obviously frustrated by the politics of the debate over the barrier, had proposed routing it through an alley behind the luxury lofts. The new route would leave about thirty-five buildings—some of the most expensive real estate in Hoboken—exposed to rising waters and storm surges. "If they don't want to be part of this, they can take care of themselves," Zimmer told me.

In some cases, walls just make water problems worse. Half a world away, in Bangladesh, building walls and embankments have actually exacerbated flooding in parts of the massive Ganges-Brahmaputra Delta. To protect their land, some farmers have built embankments along tidal channels. In doing so, however, they have inadvertently channeled the water deeper inland, causing a huge increase in flooding and saltwater contamination of less-protected areas. Similarly, a wall around Lower Manhattan might actually deflect more water into places like Red Hook, said Alan Blumberg, an oceanographer at the Stevens Institute of Technology in Hoboken. "It might keep water out of Manhattan, but it could make the problem worse for people in Brooklyn, not better."

There is also the question of complacency. Walls, dikes, and levees make people feel safe, even when they are not. When Hurricane Katrina hit New Orleans, many people didn't evacuate because they assumed the levees would not fail; that assumption cost some people their lives. In 2008, when a typhoon hit China's Pearl River Delta, one third of the granite seawalls in Zhuhai crumbled, letting water flood into the city. "Walls often make people stupid," said Richard Jorissen. "They allow you to ignore the risk of living in dangerous place—if something goes wrong, it can be a catastrophe."

There were other, less brutal ideas for how to protect Lower

Manhattan. Even before Sandy hit, a team headed by New York landscape architect and urban designer Susannah Drake proposed elevating the edge of Lower Manhattan about six feet, waterproofing utilities in vaults under the sidewalks, and raising and redesigning streets to allow them to hold water during floods. The waterfront would be softened with salt marshes and wetlands to absorb wave energy and clean storm water. To finance this new infrastructure and blend the elevated city into the waterfront, Drake's plan allowed for a row of new towers to be built along the East River. All in all, it was an elegant reimagining of Lower Manhattan in a world of rising waters. But projects like this are nuanced and complex and expensive, making them difficult to sell as a quick fix. And they require people to acknowledge that the world is changing fast and that they will live differently in the future. It's so much easier to just build a wall and forget about it—"until a big storm comes along and washes away the wall," Drake said. "Then you have a disaster."

One of the most innovative proposals to come out of the post-Sandy Rebuild by Design competition was called Living Breakwaters, which received $60 million in federal funding. The project was designed by SCAPE, a design firm in New York City founded by Kate Orff, who gained notoriety a few years ago with a bold proposal to clean up New York City's harbor by reintroducing oysters. Her Living Breakwaters project, which will be built on the south shore of Staten Island near the town of Tottenville, is a four-thousand-foot-long system of breakwaters located about a thousand feet from shore. It is not designed to stop sea-level rise. It is designed to slow and soften waves before they hit the coast, lessening the impact of storms and slowing erosion. The breakwaters themselves will be built with ecological features like textured concrete units that make healthy habitats for young fish. They will also be seeded with oysters to help further slow and clean the water. Instead of cutting the community off from the waterfront with a fortresslike wall ("the era of big infrastructure is over," Orff says bluntly), Orff wants to engage the community with the coastline again. Among other things, SCAPE hopes to work with schools to help garden oysters and start collection programs for shells that could be added to the breakwater to strengthen its growing ecosystem. "Sustainability and resiliency can be built, but by reconnecting with our shorelines, not walling off the 500-plus miles of the city's coastline," Orff wrote in the *New York Times*.

Perhaps the boldest proposal for protecting the city was the Blue Dunes, a forty-mile chain of islands that a group of scientists and architects proposed building in the shallow water about ten miles off the coast. From the city, the dunes would have been invisible, but together they would have formed a protective necklace of sand running from Staten Island up to Long Island. Like SCAPE's Living Breakwaters, the Blue Dunes were designed to absorb the wave energy of the Atlantic before it hits the city, lower the impact of high tides, and buy the city time to recalibrate for sea-level rise. But whereas Living Breakwaters is modest in ambition and human in scale, the Blue Dunes, proposed by a group headed by Dutch landscape architect Adriaan Geuze, would have reshaped the entire coastline of New York City. The Blue Dunes would not save the city from sea-level rise, but they might save New Yorkers from *fearing* sea-level rise, showing them that there are ways, as Geuze has put it, of "working with nature, bending its will, rather than trying to punish it."

The Blue Dunes provoked a lot of discussion during the Rebuild by Design competition, but in the end, the project was not funded.

One of the Blue Dunes, a forty-mile-long stretch of artificial islands designed to absorb wave energy before it hits New York. *(Illustration courtesy of West 8 and WXY)*

New York City mayor Bill de Blasio does not have a reputation as a visionary leader. But on climate change, he has a solid record, despite the fact that it's not an issue that de Blasio came to himself. It was forced on him by Sandy, which hit the city just as the mayoral election was getting under way in late 2012. Michael Bloomberg, New York's mayor at the time, had long been pushing climate change, including a landmark study called PlaNYC, a twenty-five-year plan for a greener city that he released in 2007. De Blasio, a former city councilman and political operative (he managed Hillary Clinton's New York Senate campaign in 2000), was interested in education and economic inequity. But after Sandy hit, de Blasio, who was living in Park Slope, Brooklyn, at the time, got schooled in the dangers of climate change and extreme weather. To his credit, de Blasio immediately understood that Sandy did not treat everyone equally. He told the *New York Times* a few months after the storm, "You can look at this as 'We need seawalls,' or you can look at this as 'We need to retool our approach for human security, economic security, for economic equity.'"

Rebuilding the city after Sandy was a joint city, state, and federal project. Almost all the funds came from a $60-billion federal disaster relief appropriation from Congress, which has been doled out through the US Department of Housing and Urban Development to various state and local agencies. Shaun Donovan, who was head of HUD at the time, is a native New Yorker and is widely praised for his response to Sandy. But rebuilding from Sandy is not the same as rebuilding for the city's long-term future. And in that, the city has had very little help from Washington, much less from the state capitol in Albany. New York governor Andrew Cuomo has put some muscle into greening the state's energy grid, but the reconstruction of New York City didn't earn much of his attention (within City Hall, many believed it was personal—Cuomo, who thinks of himself as the big dog in New York State Democratic politics, didn't want to do anything to make his archrival de Blasio look good). In the aftermath of Sandy, Cuomo commissioned a high-level study about how to make the state of New York more resilient to climate change—then hardly mentioned it again after it was complete. Some of his pet projects, such as a $4-billion proposal to renovate the aging LaGuardia Airport, which is in a high-risk flood zone, make no sense in a world of rapidly rising seas.

With a checked-out governor, de Blasio's leadership is all the more vital. I met with him on Earth Day in 2016, just after he made a brief speech at the United Nations to celebrate the

signing of the Paris climate agreement. In his speech, he right-
fully touted the city's progress in improving building efficiency
and purchasing more renewable power, among other CO_2-
reduction measures. De Blasio deserves a lot of credit for pushing hard to shrink the carbon footprint of New York, and he
often speaks convincingly about the implications of climate
change for the poor and working class, but I wondered if maybe
it was time for some strategic thinking about the long-term survival of the city too. Stuff like: Was it time to consider moving
the city's airports to higher ground? How about creating economic incentives to encourage people to move out of low-lying
areas of the city? If it takes New York City fifty years to construct
a single new subway line, what hope is there of rebuilding the
waterfront of the city in time to deal with rising seas?

De Blasio resisted my line of questioning, preferring to
focus on the climate challenges the city faces today and
tomorrow. "I think the simple way to think about it is right
now we have to do the most immediate resiliency measures
to secure us against the kind of storms we'd have," he told
me. "Then you want to just keep going, and building up,
building up, and trying to stay ahead of what will be a growing problem. But to me it's literally like, block by block by
block. Complete this phase and you roll immediately into the
next. This has to be a priority of government perennially
until we build a very, very different world."

I said, "When you look at flood maps that project five, six
feet of sea-level rise...it's a pretty apocalyptic scenario for
New York, isn't it?"

"Yeah. At the end of the century, true."

"That's not that long from now," I replied.

"Yes it is," he argued.

"Your grandkids will still be here."

"Yeah, but as a public policy matter, if you're talking seventy-five, eighty, or more years in the future, I think it's very, very responsible to say, 'Okay, first let's deal with the needs of people right now,' and that is both about resiliency and environmental concerns, but it's also, the totality of human need. If we don't have that in the foreground, there's something wrong with us. Right?"

True to his word, in the months after de Blasio and I talked, the city released new guidelines to encourage architects and engineers to identify design elements that could reduce the risk of flooding from future sea-level rise, including elevating new buildings by as much as three feet and erecting site-specific flood barriers. The guidelines are not part of the city's building code yet, but perhaps in a few years they will be. That was yet another step down a long and increasingly wet road.

Klaus Jacob was Hurricane Sandy's Cassandra. Jacob is a retired research scientist at Columbia University's Lamont-Doherty Earth Observatory, where he spent forty years researching earthquake prediction science and disaster relief. For the last decade or so, he had been deeply involved in shaping New York's response to rising seas as a member of the city's task force on climate change. At eighty, he still spoke with a hint of a German accent and had a bratty twinkle in his eye (five minutes after we met, he mentioned that he used to hang out with 1960s black activist Angela Davis).

A few months before Sandy hit, a Columbia University research team headed by Jacob released a case study estimating the effects of a 100-year storm surge on New York's multibillion-dollar transportation infrastructure. Jacob told anyone who would listen that the combination of rising seas and a particularly powerful storm could wreck the city's trains and subways, flooding tunnels and submerging above-ground equipment. As it turns out, that's exactly what happened when Sandy blew through. The subways were out of commission for days, and it took weeks before a system that serves millions of commuters was fully back online. Thanks in part to Jacob's warnings, New York officials shut down the subway before Sandy arrived, limiting the damage.

Jacob was critical of de Blasio and others for not thinking big enough about what kind of future the city faces. "They are thinking on an election time scale," Jacob said. He cited the continued development of waterfront property in Manhattan, Cuomo's plans to renovate the terminals at LaGuardia Airport, and Columbia University's new Manhattanville campus, which is located on low ground on the West Side near 125th Street. "We still allow development on the waterfront to take place where fifty, eighty years from now it will be regretted," Jacob told me. Even businesses that should know better are failing to grasp what's coming. Jacob pointed out that Con Edison, the utility that powers most of the city, proposed spending $1 billion on rebuilding after Sandy without taking climate change into account (that changed after ratepayers filed suit against the company; Jacob was a technical consultant in the case).

In Jacob's view, New York's Achilles' heel is the subways, which are particularly vulnerable because the tunnels need fresh air for ventilation and can't be sealed off. And although the subway tunnels are designed to handle flooding from freshwater, salt water is highly corrosive to electrical circuits, as well as to the concrete in the tunnels (that's a big reason why the L train will be shut down for more than a year of repairs). In theory, the ventilation ducts can be raised, and barriers can be erected to keep seawater from pouring in during storms, but at some point, the cost gets prohibitive.

"It's all about money," former Port Authority chief Chris Ward told me. Ward pointed out that the Metropolitan Transit Authority, which operates the New York subways, spent $530 million upgrading the South Ferry station in Lower Manhattan after the old one was heavily damaged during the attacks on September 11, 2001. After Sandy turned the new station into a fish tank, the MTA put the old station back into service while it spent $600 million to fix the new station. The MTA has installed removable barriers to stop seawater from flooding the new station in the next big storm, but the subway system remains highly vulnerable to rising seas. "We're not thinking systemically about climate change," said Michael Gerrard, director of the Sabin Center for Climate Change Law at Columbia Law School. "It's not just about the next Sandy, as if Sandy were the worst thing that could happen."

Back in the 1920s, developers discovered a spit of land out in Jamaica Bay where cops and firemen and plumbers and other working-class folks from Manhattan had set up weekend fishing camps. The shallow bay in Queens was full of crabs and bluefish and bass. Occasionally a dead horse might float by, dumped in the bay by someone who didn't want to pay to have the animal hauled off, but nobody thought much about it. Many of the structures on the sandy island were just wooden shacks that would wash away whenever a big storm hit. The people who owned them didn't care, they just hammered together a new shack and went fishing again. It wasn't much different from the way the Calusa lived in South Florida with their *palapas* and shell middens five hundred years ago.

Before long, developers came and dredged the channels in the island, thinking it would add to the appeal of the place if homeowners could dock their boats in their backyards. Then they put up houses, real houses with concrete foundations and insulated walls and septic systems. They gave the place a name—Broad Channel. Many of the cops and firemen gave up their shacks and settled into the new homes as permanent residents. They had a view of Manhattan and could go fishing after work. What was not to like? They knew they might get hit by an occasional storm but believed that the Rockaways—a much bigger barrier island to the east—would protect them from the worst of it. As for rising seas—back then, that was something that happened only in science fiction novels.

One family who bought a house in Broad Channel in the 1920s was the Mundys. When I visited in 2016, Daniel Mundy, who had been born in Broad Channel and was now in his late seventies, lived in a house right on the water. His son, Daniel Mundy Jr., who was in his early fifties, lived across the canal, also facing the water. When Hurricane Sandy hit, they were lucky— they got five feet of water in their homes, but the houses suffered little structural damage. Others were not so lucky. Broad Channel was one of the most heavily damaged areas of the city—more than 1,200 homes were flooded; over 400 needed to be torn down and completely rebuilt. Others have been repaired and elevated.

"We were really hammered," Dan Mundy Jr. said, standing in the living room on a sunny day, the towers of Manhattan glinting in the distance. Mundy was a battalion chief with the New York City Fire Department, an earnest, muscular guy who knew how to pull the levers and switches in the New York political system.

Standing on his rear deck, Mundy pointed at two narrow islands that he and his friends in Broad Channel built out in the bay (with the help of the US Army Corps of Engineers) to help bring down nitrogen levels and improve fish habitat.

Thanks in part to the work of people like Mundy, one of the most important bird estuaries on the East Coast and a spawning ground for horseshoe crabs is making a comeback.

Mundy was well aware of the risks of sea-level rise. I asked him if he ever thought about leaving.

"Both my parents were born here," he said bluntly. "I grew up here. My sister lives here. I've scuba dived under the piers in the bay five hundred times. Why would I leave? This is my home."

Many others I talked to in Broad Channel felt the same way. They were staying there come hell or high water. And the Army Corps of Engineers helped perpetuate the idea that sticking around was a viable alternative by fast-tracking a protection plan for the entire Jamaica Bay. The $2-billion project would reinforce the dunes that face the ocean on Rockaway Beach, raise seawalls in the bay, and erect a movable barrier across the inlet that could close to protect from a storm surge. Basically, it was a dumbed-down version of the MOSE barrier in Venice. And it had all the same problems: it was hugely expensive, it would take decades to build, and by the time it was built, it could very well be obsolete. But if you wanted to buy the people who lived around Jamaica Bay a little more time, there weren't many other options. Some neighborhoods around the bay, such as Howard Beach in Queens, were particularly problematic, with rows of brick homes with basements built in low-lying areas that are virtually impossible to elevate.

Later that night, I went to a community meeting, where Dan Falt, the project manager for the Army Corps of Engineers, laid out the gist of the plan for the first time to members of the community. Falt, who lives in a bungalow on Rockaway Beach, worried that the plan was going to be controversial—even though, strictly speaking, the plan was nothing new. The Army Corps of Engineers had proposed a nearly identical plan to protect the bay back in 1964; they just made a few tweaks for 2016.

But if this meeting was any indication, people who live around Jamaica Bay didn't care about how old the plan was. They wanted protection. Or at least, the illusion of protection. And they believed it was the government's role to provide it for them. Who cared if the barrier turned the bay into a big pond? The basic view of people who spoke up at the meeting could be summed up like this: It's nice to have birds and marine life, but it's even nicer to have a home—if only for a few more decades. There was no talk about living with water, about elevating homes, about commuting by boat. Nor did anyone entertain the idea that maybe the smart thing to do was to get out while they could and move to higher ground. As one woman said, standing up in the back of the room with a hint of panic in her voice, "What I want to know is, how long is it going to take to build these walls? How long are we going to be vulnerable? One more big flood, and we are toast."

An hour later, I walked back to my car, which was parked on the street near Mundy's house. It was high tide—and the street was flooded. The full moon hung above me, indifferent to the complications its gravitational pull was causing for New Yorkers. I took off my shoes and socks, rolled up my pant legs, and waded through the cold, briny Atlantic to my car.

8. ISLAND STATES

The hall at Le Bourget, an old airport on the outskirts of Paris, smelled of wet plywood and cologne. The cavernous room was one of several at Le Bourget that had been remodeled for the Paris climate talks, and now, on the final night of the negotiations, it was packed with people from virtually every nation in the world, a sea of black and brown and white faces, some dressed in business attire, some in ceremonial robes and gowns. Standing among them, I watched French president François Hollande stand at the podium in the front of the hall and slam the gavel down, marking formal acceptance of the 2015 Paris climate accord. Everyone cheered. Some cried. To my surprise, I hugged the person next to me, a young Asian woman I had never met or talked to. I felt her pull away, perhaps shocked that a stranger would grab her so suddenly, but then she hugged me back. I never learned her name or even what country she represented, but our shared expression of the power of the moment was genuine.

The Paris accord was a complex agreement, full of nuance about voluntary emissions targets and how rich nations would help finance clean energy in poor nations. But the gist of the deal was that virtually every nation in the world vowed to reduce greenhouse gases in the coming decades so as to limit the warming of the climate to 2 degrees Celsius above preindustrial levels, the well-established threshold that might allow us to escape the worst impacts of climate change. Even before President Trump reneged on America's commitment to the deal, it was unclear whether the agreement was just a feel-good moment that everyone would soon forget or whether it marked the beginning of a serious effort to bend down the trajectory of carbon pollution in the coming decades. However it turns out, one of the unsung heroes of the talks was Tony de Brum, the foreign minister of the Republic of the Marshall Islands.

De Brum, who was seventy at the time of the negotiations, was everywhere at Le Bourget, his red-and-blue-striped tie flopping over his shoulder as he shuffled quickly between meetings, often flanked by a consultant from Independent Diplomat, a nonprofit advisory group that was working with de Brum during the negotiations. With graying hair and glasses sliding down his nose, he had the air of a distracted chemistry professor. But de Brum also projected a seriousness about his mission in Paris that was hard to ignore. For people of the Marshall Islands, as well as other low-lying nations of the world, these climate negotiations were not about economic competitiveness or a global power play. They were about life and death. "My country is facing extinction," de Brum told me. De Brum was largely responsible for putting together what became known as the "high ambition" coalition during the Paris negotiations—a coalition that eventually included the United States, as well as a number of small island states like Bermuda and the Maldives—which successfully pushed for an "aspirational" goal of holding warming to just 1.5 Celsius above preindustrial, as well as zero carbon emissions by 2050.

Marshall Islands foreign minister Tony de Brum entering the hall during the final night of the Paris climate talks. *(Photo courtesy of IISD/Kiara Worth [http://www.iisd.ca/climate/cop21/enb/images/11dec/3K1A4285.jpg])*

De Brum's moral standing in the negotiations was indisputable. And it wasn't simply because the Marshall Islands would suffer a disproportionate share of the impacts of climate change. Few places in the world were treated as badly by the twentieth century as the Marshall Islands. De Brum had borne witness to it. At age nine, he was fishing with his grandfather when he saw a flash—from one of the American atomic bomb tests—on the horizon. "Within seconds, the entire sky had turned red, like a fishbowl had been put over my head and blood poured over it," he recalled. Between 1946 and 1958, the United States conducted nuclear weapons tests on the islands, peppering Bikini Atoll alone with twenty-three bombs. The largest, known as the Bravo shot, was a thousand times more powerful than the Hiroshima bomb and vaporized three small islands.

For years, the US argued that no Marshallese had been hurt during the nuclear testing. But that was not true. Bikini was evacuated, but the wind blew radioactive detritus onto the atolls nearby. "Within hours, the atoll was covered with a fine, white, powderlike substance," recalled Jeton Anjain, who led the eventual evacuation of Rongelap, one of the atolls most affected. "No one knew it was radioactive fallout. The children played in the snow. They ate it." Cancers, particularly of the thyroid, later riddled many of those who came into contact with this radioactivity.

Now rising seas, thanks largely to carbon pollution from the rich nations of the world, are threatening to extinguish life and culture in the Marshall Islands, as well as in many other poor and low-lying nations that did nothing to contribute to their own demise.

"For us, the end has already begun," de Brum told me shortly after the conference ended. "The question is, what are we—and what is the world—going to do about it?" Sadly, de Brum never learned the answer to that question. He died in the Marshall Islands in 2017. Hilde Heine, the president of the Republic of the Marshall Islands, called him "a national hero."

Majuro, the capital of the Marshall Islands, is an urban oasis in the Pacific: about 30,000 people live on a thin spit of land that encircles a mile-long lagoon. The moment you get off the plane, the ocean is everywhere, suffusing the air with salt and hinting at the unimaginable vastness of the surrounding waterworld. Life is precarious here, and you don't have to be a climate scientist to see that in a world of quickly rising seas, this place is in big trouble.

The Republic of the Marshall Islands is already far more water than solid ground. It's made up of about 70 square miles of land, which are spread over more than 1,000 islands and atolls in 750,000 square miles of the Pacific Ocean between the Philippines and Hawaii. The "land" that makes up the atolls was created from coral reefs that grow on the submerged remains of long-extinct volcanos. Most of the atolls are five or six feet above sea level, although a few spots on Majuro are as high as twenty feet. Other island nations like Kiribati, which is south of the Marshalls, and the Maldives, which are in the Indian Ocean, have a similar low, flat topography. And, in a world of quickly rising seas, a similarly dark future.

Of course, so do lots of other places. But what makes the predicament of the island states so tragic—and why Tony de Brum had so much authority in Paris—is two things. First, they are suffering because of the indulgences of others. Climate change was set in motion because of the two-hundred-year-long fossil fuel party the West has been throwing for itself (as I mentioned in an earlier chapter, CO_2 accumulates in the atmosphere—so if you want to know who is responsible for our warming atmosphere, you have to look at historical emissions). To put in perspective how little the Marshallese have done to cause the problem, consider this: the entire amount of CO_2 emitted by the Marshallese in the last 50 years is less than the city of Portland, Oregon, emits in a single year. And of course, the basic injustice of climate change is that the people who are least responsible for the problem are the ones who will pay the most dearly for it. The most pressing question that rich Western nations have faced during the last thirty years of climate negotiations, and will probably have to face in the next thirty years of climate negotiations, is simply this: *What do we owe them?*

But for island nations, this question is particularly urgent because what is at stake for them is not just their homes and their livelihoods, but also their language, their culture, their identity. There is no high ground to migrate to in the Marshalls; when they go under, everything is gone. As de Brum put it in an interview shortly before the Paris talks began, "Displacement of populations and destruction of cultural language and tradition is equivalent in our minds to genocide."

In the decades since the US Department of Defense turned the islands into a testing ground for nuclear weapons, poisoning the land and the people, the Marshallese have received billions of dollars from the United States in various subsidies and payments. Wherever that money has gone, it has clearly not made up for the psychological and cultural devastation of being the kitty litter box for the Nuclear Age.

This is a country that uses faded US dollar bills and has no economy to speak of. The country produces coconuts and breadfruit and sells fishing licenses for the international trawlers that scour Marshallese waters for tuna, but it leans heavily upon the aid of others. Majuro's police station, courts, and street lighting were funded by Japan, Taiwan, and Australia. The lack of a modern sewer system means contaminants are destroying the lagoon that surrounds the atoll, killing fish and making fishermen go farther and farther to find their catch, wasting fuel.

Majuro is also poorly positioned for battle against the attacking sea. It's located on narrow windward islets, a carry-over from the days of military occupation when the city sprang up around an American seaplane base. Before that, people had more prudently lived on the much wider island on the leeward side of the atoll. Seawalls are in bad repair, and there is no money to build new ones. Star architects like Rem Koolhaas are not constructing modern, sea-level-rise-worthy buildings here. Almost everything is cinder block and tin roofs. The only resilient structures are the boats that often drop anchor out in the lagoon (*Senses*, a 193-foot yacht owned by Google cofounder Larry Page, has been spotted offshore).

To see what a fortress on the sea looks like, Marshallese have only to take a short boat ride out to Kwajalein, the largest of the Marshall atolls, which is home to the Ronald Reagan Ballistic Missile Defense Test Site. The twelve hundred Americans who live on the base launch missiles, operate space weapons programs, and track NASA research, supported by an annual budget of $182 million.

Kwajalein is not immune to the threat of rising seas. But the US military has no problem finding the money for new seawalls and other protective infrastructure. In 2008, when a storm surge flooded the base and destroyed all the freshwater supplies on the island, the military responded with expensive desalination machines and heavy-duty barriers made of a fortified granite commonly called riprap. The Pentagon is apparently feeling confident enough about these fortifications to invest $1 billion in a new radar installation on the atoll designed to help satellites and astronauts avoid colliding with space junk as they orbit the Earth.

"It is a different world out on the base," one Marshall Islander who works at the test site told me. "It is a place where you feel safe."

Outside the base, there are only sand and crumbling seawalls and water as far as you can see. The questions that view provokes in the Marshallese are elemental: How long can I stay? And when it's time to leave, where do I go?

On coral atolls like the Marshalls, there are no spring-fed rivers, no mountain lakes, no bucolic streams. Freshwater comes from the sky. The Marshallese collect water in rain buckets on the roofs of their houses, or in collection ponds at the airport. Mother Nature sometimes collects it in narrow underground aquifers—what geologists call freshwater lenses. As long as the rains continue, everything is okay. But if it stops raining for a few months, as it did in 2013, and again in 2015, then there is trouble. Marshallese find themselves in the very Robinson Crusoe–like predicament of dying of thirst while surrounded by water.

Freshwater has always been an issue on coral atolls. But as populations have grown, the problem has become more acute. It's one thing to collect enough rainwater on an island for three hundred people. It's something else entirely to do it for a city of thirty thousand.

The Marshallese have dealt with this in a sensible way—by increasing the amount of water they can collect. But there is not enough land to build enormous reservoirs; you also can't dig too deep before you hit salt water bubbling up from below. Instead, they have built a series of collection ponds near the airport to capture runoff from the tarmac—they look like a row of swimming pools covered with black plastic (to reduce evaporation). The water is polluted by grease and oil and bird droppings, so it has to be treated and filtered before it can be drunk. When the reservoirs are full, they hold 34 million gallons of freshwater, enough for the people of Majuro to survive for several months.

Only about one quarter of the people on Majuro are connected to the municipal water supply from the airport, however. The other twenty-two thousand or so residents—as well as another twenty thousand people who live scattered around on outer atolls—rely on rainfall either collected in plastic drums or pumped out of the freshwater lens in the ground. By necessity, the Marshallese are very good at conserving fresh water. In New York City, the average person uses 118 gallons per day. In Majuro, it's 14 gallons per day.

Nevertheless, running out of drinking water is a perpetual worry in the Marshalls. The northern atolls get less than fifty inches of rain a year. Atolls to the south get about double that. As the climate warms, rainfall patterns are likely to change. According to some models, the Marshall Islands on average will get more rain in the coming decades, not less—and they'll also get hotter temperatures and longer droughts.

But whatever happens with rainfall patterns, drinking water is likely to remain a concern. The airport reservoirs are surrounded by dikes, but nevertheless, when a big storm comes in, the waves sometimes break over the walls. "If we get overtopping, where salt water gets into our catchment, we have to dump the freshwater out into the lagoon," Kino Kabua, the deputy chief secretary in Majuro, told me. "As climate change comes in and sea levels rise, it's going to increase that danger."

The larger threat comes from salt water infiltrating the many lenses of freshwater on islands and atolls. These lenses are extremely sensitive and constantly shifting. They are often depleted during the dry season, becoming so brackish that they are undrinkable. When rain finally comes, it replenishes the lens, filtering down through the sandy soil and into the porous coral rock that forms the base of the atoll. The thickness of the lens depends on the size of the island, the amount of rainfall, and the height of the ocean. As you go deeper, the water gets more and more saline. Eventually, it becomes pure seawater.

The problem is, as seas rise, the salt water pushes up from below, leaving less and less room for freshwater (which, being more buoyant, rides on top of the salt water). In addition, as seas rise, flooding from storm surges is likely to become more common. When an atoll is inundated, the salt water can seep into the freshwater lens, contaminating it. It can take years before it is suitable for drinking again.

A parallel problem, and one that is equally serious, is soil salinization. Tony de Brum often talked about the fact that it's getting harder and harder to grow breadfruit, a staple food in the Marshall Islands, because the trees will no longer tolerate the salty soils. In the Marshalls, the soil contamination comes from using salty water for irrigation, but also from the increasingly frequent flooding that occurs during high tides and storm events. All over Majuro, you can see wilted and dying breadfruit trees. Other staple crops, such as bananas and papayas and mangoes, have also been affected.

Government-run farms are experimenting with salt-resistant hybrids of crops such as taro and cassava. But as the soil gets saltier, the Marshallese will become more and more dependent on food imports. Even now, the old island diet of fish and fruit is mostly gone. Today, rice, flour, and meat are staples for breakfast, lunch, and dinner. Almost all of it is imported—at high prices (the Marshall Islands produce no rice, no wheat, no cattle). And the Marshallese are bearing the costs in other ways, too: 65 percent are overweight or obese. More than 30 percent have diabetes, one of the highest rates in the world.

Increasingly salty soils and drinking water are not just a concern for island states. In Miami, salt water is gradually pushing deeper in the shallow aquifers beneath the city, threatening the region's drinking water supply. In Vietnam's Mekong Delta, salt-water-contaminated soil is rendering once-productive land sterile, leaving millions of people without traditional crops like rice. In Egypt, freshwater supplies are decreasing at alarming rates, and the country may face nationwide shortages by 2025. According to one recent study, dwindling access to freshwater and the increasing salinity of the Nile Delta's agricultural land threaten to make the country uninhabitable by 2100. In Bangladesh, a study by the World Bank and Bangladesh's Institute of Water Modelling published in 2015 suggests that in a worst-case scenario, the area served by freshwater rivers in the country's coastal areas will drop by half in the coming decades. "The biggest problem we have is with water," Asma Begum, a villager in Gabura on the coast of Bangladesh, told a BBC reporter. "It is everywhere, but it is not drinkable. And it is destroying our land."

Salt is also destroying Bangladesh's rice crop, just as it is in Vietnam and other delta regions in South Asia and Southeast Asia. This has tremendous implications for food security, since rice accounts for 70 percent of the calories consumed by the country's 160 million people. Researchers are experimenting with more salt-tolerant varieties of rice, but progress has been slow. Instead, many Bangladeshis are turning to shrimp farming, since shrimp thrive in brackish waters. But shrimp farming, in many cases, just makes the problem of salinization worse. The shrimp industry has coerced farmers to transform more and more freshwater ponds into saline, which has allowed the salt water to flow deeper and deeper into the delta region, turning what was once a habitable region where families could grow vegetables and raise chickens into a practically marine environment.

In contrast to salty soils, salty drinking water has a relatively easy technological fix—if you have the money. Right now, the biggest desalinization plant in the world is in San Diego. It cost $1 billion to design and build. The plant takes in 100 million gallons of Pacific Ocean water a day and produces 54 million gallons of fresh, drinkable water. That's only about 10 percent of what San Diego County needs, but it is reliable and drought-proof. Like most other desalinization plants, this one uses a process called reverse osmosis, which forces the seawater through a thin membrane, removing the salt and other impurities. One reason it is so expensive is the tremendous amount of energy that is required to push the water through the membrane—the San Diego plant uses about 35 megawatts of power, which costs about $30 million a year. The high price of desalinization is one reason why 70 percent of the plants in the world are in rich oil nations of the Middle East. Saudi Arabia alone is planning to spend $28 billion constructing new plants in the coming years.

Poor nations like the Marshall Islands use small, portable desalinization machines for emergency drought relief. But they only produce a few thousand gallons of water a day, require a lot of fuel, and break down often. Water experts who visit the islands are experimenting with simpler technology, including plastic tents that create freshwater by evaporation. Those work well enough, although the quantities of water they produce are small and they are vulnerable to being wiped out in storms. Desalinization technology will certainly get better, cheaper, and more rugged over time. But it still introduces a technological dependency that increases the cost, the complexity, and the risk of living on a remote atoll. Some may manage it just fine. But for others, it is likely to be one more reason to leave.

On November 13, 2015, two weeks before the climate talks in Paris began, Islamic terrorists launched a wave of mass shootings and suicide bombings in the city that left 130 dead and almost 400 wounded. The worst of the attacks occurred at an Eagles of Death Metal concert at the Bataclan, a theater in Paris's Eleventh Arrondissement, where 90 people died. During the climate talks, the city was on high alert, and security was everywhere. I saw military officers with machine guns searching through the bushes near the entrance to the Louvre and armored personnel carriers in the Place de la Concorde.

It's hard to know how the attacks affected the negotiations in Paris. They certainly darkened the mood at Le Bourget and underscored the feeling that the world was becoming a more dangerous place. In retrospect, the Paris attacks marked the moment when Europe—and the world—began to turn inward, to erect barriers and walls. At the time, European leaders were already struggling to deal with the flood of refugees from the war in Syria. Although all the terrorists who participated in the Paris attacks turned out to be

EU citizens, the unspoken fear during the climate talks was that these attacks were a preview of things to come. Globally, more than one billion people live in what demographers call low-elevation coastal zones. A fair percentage of these people are in poor countries with little money to help with adaptation. As the waters rise, where will they go? What are their legal rights? What do rich industrialized nations like the United States and the European Union owe them? One way to view the past thirty years of climate talks is as an extended attempt by poor nations to extract compensation from rich nations for stealing their future.

Not surprisingly, it hasn't gone too well for poor nations. Various special funds have been set up by the UN over the years, but they were stifled by bureaucracy and the West never contributed much. After the Copenhagen talks in 2009, the Western nations set up a new mechanism called the Green Climate Fund, which vowed to mobilize $100 billion a year in public and private finance by 2020 to help developing nations deal with the problem. In Paris, the United States pledged to commit $3 billion to the fund, but by the time Obama left office had only made good on $1 billion, including $500 million Obama transferred into the fund just days before he departed. Still, the fund was woefully short of its $100 billion target—as of the end of 2016, it had about $10 billion in pledges and had paid out about $1 billion in grants. That might sound like a lot, but when you consider that the point of the Green Climate Fund is to help poor nations of the world move to clean energy, protect themselves from rising seas, secure food and water supplies, and generally brace for the future, $1 billion is hardly even a drop in the bucket. New York City will spend nearly that much just to build a wall along the Lower East Side of Manhattan.

For the Marshall Islands, the Green Fund has been little help. The total amount of money the nation has received from the fund to help with climate adaptation is close to zero (as of this writing, there are a few small projects in the works). The nation has received funding from other agencies, including the World Bank, which has given the Marshall Islands $6 million to help with adaptation projects, including $1.5 million to make buildings more resilient and hire staff for disaster preparedness. The Marshall Islands have also received funds from a number of other agencies, including the Asian Development Bank. But the simple truth is, the Marshall Islands don't have the money to protect themselves from the consequences of rising seas. Their survival depends on the largesse of others.

Some island nations have figured out other ways to raise funds. The Maldives, for example, have a much larger population (350,000 versus 50,000) and a much larger economy ($2.3 billion GDP versus $190 million GDP), largely thanks to well-developed high-end tourism. To raise money, they're reclaiming land from the sea, essentially building new islands inside existing lagoons. In 2016, the Maldives parliament passed a constitutional amendment that for the first time allowed foreign ownership of Maldives territory. Specifically, the amendment allows foreigners who invest over $1 billion to own land, provided that at least 70 percent of the land is reclaimed from the sea. Indian officials are concerned that this is an invitation to the Chinese, who have already launched a massive land reclamation project in the South China Sea. Unnamed Indian officials said they are "concerned" that

China now plans to do the same in some of the Maldives' 1,200 islands, which are located strategically in the Indian Ocean. Eva Abdulla, one of just fourteen parliamentarians in the Maldives who voted against the amendment, said, "This will make the country a Chinese colony."

During the Paris climate talks, one member of the US negotiating team told me, the Chinese were "very closely attuned" to the issues facing the small island states. This was no surprise, given the clear interests the Chinese have in expanding their presence in the Pacific. But it also gave the small island states extra leverage in the negotiations. "You can help us and be our benefactor," one island state negotiator said to a member of the US team, "or we can go talk to our Chinese friends."

In 2007, the United Nations Framework Convention on Climate Change (UNFCCC) agreed to consider adding "loss and damage" as a new category in the climate negotiations. That led to something called the Warsaw International Mechanism for Loss and Damage, which was essentially a UN committee convened to figure out how to define "loss and damage" and how to handle liability and compensation. Could big polluters like the United States be held legally responsible for the flooding of the Marshall Islands? For the lost value of real estate? How about compensation for causing a species to go extinct? If a person was killed in a flood, could his or her family sue the UK for the 150 years of carbon pollution that caused the water to come in?

It's not hard to see why the United States and other rich nations were not eager to participate in this. For one thing, it could create yet another financial obligation to the developing world at a time when most wealthy developed nations hadn't done much to fulfill earlier obligations. For another, it threatened to open the door to complex and potentially expensive lawsuits, in which a Pacific Islander could sue the United States for knowingly dumping the CO_2 into the atmosphere that caused the rising of the seas that washed away his or her house. The idea of this petrified rich nations—not because they thought it would happen anytime soon, but because it would set a precedent for legal responsibility for damages in poor nations caused by climate change. "We're not against [considering loss and damage]," Secretary of State John Kerry told me a few weeks before the Paris talks. "We're in favor of framing it in a way that doesn't create a legal remedy, because Congress will never buy into an agreement that has something like that... the impact of it would be to kill the deal."

In Paris, loss and damage was one of the most volatile issues in the negotiations. It was broadly agreed that developing nations had a moral right to compensation for the damages they suffered from climate change. There were many predictable attempts to sabotage any deal (hilariously, negotiators for Saudi Arabia, which ranks as the world's fifteenth-largest economy, demanded that if islands like Kiribati or the Marshalls were to be compensated for damages, the Saudis should also be protected from loss of future oil income). In the end, negotiators agreed only to increase "understanding, action, and support" for loss and damage associated with climate change. In other words, they agreed to keep talking.

But as the terrorist attacks in Paris reminded everyone, this was about more than just money. It was also about displaced people. Every year, about 20 million people are made homeless by floods, storms, and other disasters—roughly three times as many as those displaced by conflict or violence, according to a report by the Internal Displacement Monitoring Centre, an independent organization that provides information and analysis of the world's displaced people. No one knows for sure, of course, how many people will be forced to migrate in the decades ahead. Forecasts range from 25 million up to 1 billion by midcentury. The most credible estimate may be from the International Organization for Migration, which projects about 200 million climate refugees by 2050.

Fleeing high water on the Pacific island of Kiribati. *(Photo courtesy of Getty Images)*

These displaced people have no legal status or protection. Under international law, there is no such thing as a climate refugee. The 1951 Refugee Convention defines a refugee narrowly as someone who "has a well-founded fear of persecution because of his/her race, religion, nationality, membership in a particular social group or political opinion." But even if a specific designation did exist, capturing the range of problems that cause people to move would be tough. For Pacific Islanders, the issue is rising sea levels; for others, it's a conflict over land or water. Groups of academics, activists, and government officials have proposed nonbinding guidelines for the treatment of people who cross borders as a result of climate change, but these do not have the force of law.

Before the Paris meeting, there had been a move to create a group within the UN that would recognize and help with relocations, as well as provide compensation. But Australian negotiators killed that. During the Paris talks, the whole issue of refugees was so politically explosive that it was hardly discussed.

It was a chilly November day almost a year after the Paris talks ended. The world had changed: the United States had just elected a president who believed climate change was a hoax and would soon fill his cabinet with climate change deniers of every stripe. Benetick Kabua Maddison, a twenty-one-year-old Marshallese, was not happy about this. He stood on a busy street corner in Springdale, Arkansas, with a crowd of college students—some Marshallese, some hillbilly white—who were waving signs that said LOYAL TO THE SOIL — WE NEED WATER NOT OIL and IF WE DESTROY CREATION, CREATION WILL DESTROY US. Maddison had a scruffy beard and wore black-framed glasses and a black beanie and walked with a slow saunter. You would never know it by looking at him, but he comes from a royal lineage in the Marshall Islands (Amata Kabua, the first president of the Marshalls, was his grandfather's cousin). Maddison was born on Majuro and lived there until he was six years old, when his family decided to move to the United States. Today he studies political science at Northwest Arkansas Community College and lives in a four-bedroom house in Springdale with his parents and seven brothers and sisters. After he finishes college, Maddison expects to return to the Marshalls, get involved in politics, and someday become the island's president.

The street-corner demonstration was a protest against Trump, but more important, it was a way of announcing to the people of Springdale (and the president-elect) that issues like immigration and climate change mattered to them. Maddison and his family live in the United States under an agreement called the Compact of Free Association, which grants the US military the right to use the atolls for whatever they want, while allowing Marshallese to live and work in the States indefinitely. They can't vote, but they can come and go freely and pay taxes and get benefits. Today, about fourteen thousand Marshallese live in Springdale, of all places — they began coming here in the 1990s, attracted by jobs processing chickens in a region proudly known as the poultry capital of the world. Tyson Foods, a $23-billion company headquartered in Springdale, slaughters about 33 million chickens each week. In the processing plants, employees face brutal conditions, working twelve-hour shifts and cutting up and pulling apart an average of about fifty chickens a minute. In one study, 86 percent of workers reported hand and wrist pain, swelling, numbness, or the inability to close their hands. Human Rights Watch found that poultry workers frequently develop "claw hand," in which their fingers get locked in a curled position. And then there are more basic indignities. "Workers in the processing plants often don't get bathroom breaks and have to pee on themselves," Maddison told me when we talked after the demonstration. "For Marshallese women, who take great pride in their personal cleanliness, this is very difficult."

The crowd of protesters walked over to a podium in the park, and Maddison stood up and gave a short speech. It was about the urgency of dealing with climate change, but it also addressed the strangeness of being a Marshallese in Arkansas. "You can't move a culture from one place to another, because land is a focal part of Marshallese culture," he told the crowd. "Just as our islands are important to us Marshallese, so is our culture. It is our identity and a gift from God bestowed to our ancestors to pass down to the ones alive today, their descendants. The Marshall Islands is like an alele bag, which in the ancient days was used to hold one's valuable possessions. If we lose the islands, then we also lose all those possessions that have been ours for millennia."

Not long after the demonstration, Maddison and I went out for lunch at a nearby diner. The walls were covered with kitschy photos and paintings of wooden fence posts and weathered barns. Maddison talked about his life in Arkansas, the church he attends, his seven brothers and sisters. His father had worked at a tool manufacturer for a few years after they arrived in Springdale, then had to retire with health problems. Besides attending classes full-time, Maddison worked with the Marshallese Educational Initiative, a community group that tutored middle school kids and offered Marshallese language classes.

I asked him if he considered himself a climate refugee. "I do," he told me. One of the reasons his family left the Marshalls fifteen years ago was the increasing flooding during high tides. He remembered one particular flood when he was about five years old. "The water just rushed into the house," he said, his voice quickening. "I was in it up to my waist. I remember toys, blankets scattered everywhere. My father grabbed me. It was very frightening. The entire island was flooded."

Maddison was aware of his privileged position in the United States. "I am going to stay here and get my education. I am grateful to America for this. But I want to go back to my country as soon as I can."

In one sense, the Marshallese are lucky—under the Compact of Free Association, there are no restrictions on the number of Marshallese who can come to America. The entire population of the country could move to Arkansas. Other Pacific Islanders are not so fortunate. As waters rise and life on the islands gets more precarious, some of them have developed emergency backup plans to keep their nation alive. In 2014, Kiribati purchased eight square miles of private land in Fiji for $8 million. Fijian president Ratu Epeli Nailatikau has assured the people of Kiribati that some or all of them would be welcome to migrate thirteen hundred miles across open ocean waters to his country, if the need arose. "Fiji will not turn its back on our neighbors in their hour of need," he said in 2014. "In a worst-case scenario and if all else fails, you will not be refugees."

It was a singular statement. For climate refugees, the problem is not just that they lack legal status under international law. It's that nations of the world are turning their backs on displaced people of any sort. And it's not just Europe, where even rich, progressive nations like Denmark are throwing up barriers to keep people out. In Australia, boats containing refugees are intercepted at sea; refugees are either turned away or imprisoned on islands. And in the United States, of course, Trump won the Oval Office largely by stoking fear of dark-skinned outsiders and promising to build walls to keep them out.

A few years ago, Mohamed Nasheed, the former president of the Maldives, confronted Western polluters with the following options: "You can drastically reduce your greenhouse gas emissions so that the seas do not rise so much.... Or, when we show up on your shores in our boats, you can let us in.... Or, when we show up on your shores in our boats, you can shoot us. You pick." More likely, said Michael Gerrard, director of the Sabin Center for Climate Change Law at Columbia Law School in New York City, "we'll see the emergence of more refugee camps like in Kenya." Dadaab was a Kenyan camp that began in 1992 to offer refugee for people fleeing the war in Somalia; two decades later, it had evolved into a semipermanent tent city in the desert with a population of half a million people. "The level of misery in a place like that is unimaginable," Gerrard told me.

Gerrard argues, provocatively, that industrialized countries ought to pledge to take on a share of the displaced population equal to how much each nation has historically contributed to emissions of the greenhouse gases that are causing this crisis. According to the World Resources Institute, between 1850 and 2011, the United States was the source of 27 percent of the world's carbon dioxide emissions; the European Union, including the UK, 25 percent; China, 11 percent; Russia, 8 percent; and Japan, 4 percent. "To make calculating easy," Gerrard wrote, "let's assume that 100 million people will need new homes outside their own countries by 2050. Under a formula based on historic greenhouse gas emissions, the United States would take in 27 million people; Europe, 25 million; and so on. Even as a rough estimate, this gives a sense of the magnitude of the problem: The United

States has been granting lawful-permanent-resident status to only about 1 million people a year for several decades." Under President Trump, who made it clear during his first weeks in office that refugees of any sort will not be welcomed in the United States, that number is sure to fall precipitously.

While we ate burgers at the diner, Maddison told me that although he has not been back to the Marshalls since he left when he was six years old, he keeps in touch with family there through email and Facebook. After talking to him for a while, it became clear to me that he leads a dual life—his body is in Arkansas, but his heart and soul are in the Marshalls. When he is not in class, he spends much of his time preserving the culture of his vanishing homeland. He collects stories from elders for an oral history of the islands, and teaches Marshallese to children of Marshallese émigrés who have grown up speaking only English.

After lunch, Maddison and I walked around downtown Springdale. It was drizzling rain, and the storefronts were mostly empty. While we walked, Maddison wondered aloud about the legal implications of his nation going under. "There are a lot of questions that I don't know the answers to," he told me. "Am I still a legal resident of the Marshall Islands if the land goes underwater? What happens to our country's fishing rights?" He also talked about the Runit Dome, a concrete bunker on Enewetak Atoll where the US military buried 111,000 cubic yards of radioactive debris left behind after the nuclear tests. It sits right at sea level. "It is already cracked, and as the waters rise, it will be submerged. So we will have a problem of nuclear waste."

I asked him if he was angry at the United States for this, and for all that the US had done to his nation—for dropping nuclear bombs on his islands, for burning fossil fuels and heating up the climate and threatening his people, his nation, and his identity with, as Tony de Brum had said so bluntly in Paris, "extinction."

The future president of the Marshall Islands waited on the corner for a Walmart truck to pass, then stepped out to cross the road. "I do not think the Marshallese are angry people," he said slowly. "But we do believe in justice. If you destroy someone's country, I think you owe them something for that."

9. WEAPON OF MASS DESTRUCTION

NAVAL STATION NORFOLK is home to the US Navy's Atlantic fleet, an awesome collection of military power that is in a terrible way the crowning glory of our civilization. When I visited the base, the aircraft carrier USS *Theodore Roosevelt* was in port, a 1,000-foot-long floating war machine that was central to US military operations in Iraq and Afghanistan. The *TR*, as it's referred to on the base, was bustling with activity— cranes loaded equipment onto the deck, sailors rushed up and down the gangplanks. Security was tight everywhere. While I was out on Pier 7, one of the base's new double-decker concrete piers that is so big it feels like a shopping mall parking lot, I wandered over to have a close look at USS *Gravely,* a guided missile destroyer that has spent a lot of hours patrolling in the Persian Gulf. Armed men on the deck watched me warily—even my official escort seemed jittery ("I think we should step back a bit," he said, grabbing my arm). Navy helicopters hovered overhead, and there was a constant hum of activity as 75,000 sailors and civilians who work on the base went about the daily business of keeping the ships spit-shined and ready for deployment at any moment.

You can't spend ten minutes on the base without feeling the deep sense of history here. The Battle of Hampton Roads, a famous naval showdown between two ironclad Civil War ships, occurred just offshore. The base was a key departure point during World War II for thousands of sailors, many of whom never returned. Their ghosts still haunt the place. This is a world born of war, where everyone's aunt or uncle has a story to tell about a night in a port in Brisbane or Barcelona or about the way their ears rang the first time they heard a cannon firing from the deck of a ship. And it's a world that will soon vanish beneath the ocean. "Norfolk is the biggest navy base in the world, and it's going to have to be relocated," former vice president Al Gore told me. "It's just a question of when."

Naval Station Norfolk, the largest navy base in the United States, is highly vulnerable to sea-level rise. *(Photo courtesy of US Navy)*

Naval Station Norfolk is at risk because of a number of factors, including the subsidence of the land the base is built on and the slowdown of the Gulf Stream current, which brushes up against the coast here (as on the rest of the mid-Atlantic coast, sea levels are rising in Norfolk roughly twice as fast as the global average). All it takes is a rainstorm and a big tide, and the Atlantic invades the base—roads are submerged, entry gates impassable. When I visited the base during a nor'easter one December, there was water everywhere. It splashed over my boots as I stood at the edge of the base, looking over the gray water of Willoughby Bay. On Craney Island, the base's main refueling depot, I saw military vehicles up to their axles in seawater. It pooled in a long, flat grassy area near Admirals' Row, where naval commanders live in magnificent houses built for the 1907 Jamestown Exposition, which was held on the grounds here. There is no high ground on the base, nowhere to retreat to. It feels like a swamp that has been dredged and paved over—and that's pretty much what it is.

Norfolk—and the smaller cities nearby, sometimes known collectively as Hampton Roads—is the heartland of the US military. It's just a few hours' drive from the Pentagon. There are twenty-nine other military bases, shipyards, and installations in the Hampton Roads area, and most of them are in just as much trouble. At nearby Langley Air Force Base, home to several fighter wings and headquarters of the Air Combat Command, base commanders keep thirty thousand sandbags ready to stack around buildings and the runways so they remain usable at high tide. At Dam Neck, another navy base, they stack old Christmas trees on the beach to keep the shoreline from eroding. At NASA's Wallops Flight Facility, where satellites are launched into orbit, plans are already in motion to move the launchpads back from the beach. "Military readiness is already being impacted by sea-level rise, and it's just going to get worse," Virginia senator Tim Kaine told me. How much worse it's going to get, and how fast, Kaine would not say, in part because he doesn't want to start a stampede out of the region, and in part because no one knows for sure.

For now, the strategy is just to buy time. Since the mid-1990s, the navy has spent about $250 million to build four new double-decker piers to withstand the rising water. It would cost another $500 million or so to elevate the remaining piers at the base, but that would do nothing to save the many roads and buildings and runways on the base, all of which are critical infrastructure and all of which are in danger. But a base like Norfolk is not just barracks, piers, and ships. It is the hub of an entire ecosystem that has grown up around it during the last century—fuel suppliers and electrical lines and railroad tracks and repair shops and reasonably priced housing stock and decent schools for the children of the men and women who are stationed there. You can't just move all this to some random spot on the coast of New Hampshire. "You could move some of the ships to other bases or build new smaller bases in more protected places," said Joe Bouchard, a former commander of Naval Station Norfolk. "But the costs would be enormous. We're talking hundreds of billions of dollars."

A few months after my initial visit to the base, I returned with Secretary of State John Kerry. Kerry was visiting the base to help celebrate the 250th anniversary of the US Marine Corps aboard the USS *San Antonio,* a state-of-the-art amphibious landing ship designed to deliver up to 800 marines ashore via landing craft and helicopters. Kerry said a few words at a ceremony on the lower decks, ate some birthday cake with Marine Corps officers, then headed up to the bridge of the ship to speak to the troops on the ship's PA system.

From the bridge, Kerry had a commanding view of the base—aircraft carriers to the left, battleships to the right, a panorama of military power. After Kerry made his remarks to the troops, he got an informal briefing from navy officials about the base's vulnerability to sea-level rise. Already, roads connecting the base to the city of Norfolk flood with major rainstorms, they told him. At high tide, they continued, water surges over the seawalls, threatening key infrastructure and inundating buildings. Kerry, dressed in a sharp blue suit and pink-orange tie, asked the naval officers about the life expectancy of the base. "Twenty to fifty years," Captain J. Pat Rios told him.

There was a slight but perceptible pause among the naval officers and State Department officials on the bridge. It was an extraordinary moment in the annals of American military history: a US naval officer had just told the secretary of state that this enormous naval base, home to six aircraft carriers and key to operations in Europe and the Middle East, would be essentially inoperable in as little as twenty years. Yes, they could shore up the seawalls for a while. Yes, they could raise roads. But without the massive influx of billions of dollars to fortify and elevate the city of Norfolk, as well as the roads and railroads that connect it to the surrounding region, the base was in big trouble.

Kerry asked a few follow-up questions about what was

being done now to buy more time, but he hardly seemed perturbed. Part of the reason for that may have been that this daylong tour of the Norfolk naval base was a brief diversion from his more immediate concerns, which was trying to stop the bloodshed in Syria and figure out a way to counter Russian president Vladimir Putin's rising influence in the region. But a larger part of the reason was that the troubles at the naval base were hardly news to Kerry. He had been talking about the national security implications of climate change for years. But now, reality was starting to catch up with him.

The scale of the military assets that are at risk due to our rapidly changing climate is mind-boggling. The Pentagon manages a global real estate portfolio that includes over 555,000 facilities and 28 million acres of land—virtually all of it will be impacted by climate change in some way. And it's not just active bases and military installations that are in trouble. The headquarters of the US Southern Command, which is in charge of military operations in South and Central America as well as the Caribbean, is located in a low-lying area near Miami International Airport that is already vulnerable to flooding. The United States Naval Academy in Annapolis, Maryland, is perched right on the edge of Chesapeake Bay and is often inundated at high tide.

On the East Coast, at least four key military bases are at risk from sea-level rise and storm surges, including Eglin Air Force Base, the largest air force base in the world, which is on the Florida Panhandle. Up in Alaska, the problems are thawing permafrost and coastal erosion that is accelerating with higher tides. The air force's early-warning radar installations, which help the United States keep a close watch on anything lobbed our way from North Korea or Russia, have been hit particularly hard by coastal erosion. At one radar installation, forty feet of shoreline have been lost, endangering the reliability of the radar.

In some places, these impacts are little more than expensive nuisances. But in others, the future of entire bases is in question. And many of these bases are virtually irreplaceable because of their geography and strategic location. The missile base in the Marshall Islands that I mentioned in the previous chapter is one example. Another is the US naval base on Diego Garcia, a small coral atoll in the Indian Ocean, another strategic military asset that is already threatened by rising seas. The base, which was built during the Cold War, gave the US military a footing from which to counter Soviet influence in the region, as well as to protect shipping lanes out of the Middle East. The base has become a critical logistics hub for sending supplies to joint forces in the Middle East, the Mediterranean, and southern Europe. It also houses Air Force Satellite Control Network equipment used to control the global GPS system. The ships and equipment can be moved easily enough, but giving up a toehold in a vital but flammable part of the world is not something the military likes to do. "To the navy, presence matters," retired admiral David Titley, who now heads the Center for Solutions to Weather and Climate Risk at Pennsylvania State University, told me. But the atoll is so low-lying that, like the nearby Maldives, it is sure to vanish unless the navy wants to spend billions of dollars turning it into a fortress in the middle of the Indian Ocean.

The Pentagon has spent several years studying its 704 coastal installations and sites to determine which bases are most at risk. Eventually some tough decisions will have to be made about which ones to close, which ones to relocate, which ones to protect. But nobody in Congress—neither Democrats nor Republicans—wants to talk about it. The first indication of these decisions will likely be revealed in the next meeting of the Base Closure and Realignment Commission, which is supposed to occur in 2019—but may well be postponed. "In BRAC, all of the decisions are based on the military value of the military installation that you have," John Conger, the assistant secretary of defense who is responsible for BRAC, told me in 2015. "Will climate change affect the military value of the installation? Well, sure it will. How can it not if I have an increased flood risk at any particular location? So it affects the military value. The question is, does it dominate the equation? And I don't think it does—*yet.*"

In Norfolk, the problems are not only geographical—they are also political. Just as there are climate change hotspots, there are also climate denial hotspots—and Virginia is one of them. Several years ago, former state attorney general Ken Cuccinelli launched a witch-hunt against noted climate scientist Michael Mann, subpoenaing documents and private emails in an attempt to discredit his work. The Republican-dominated Virginia legislature has effectively banned the discussion of climate change—one legislator called sea-level rise "a left-wing term." Instead, the politically acceptable phrase in Virginia is "recurrent flooding." Governor Terry McAuliffe, the former head of the Democratic National Committee and fundraiser for President Bill Clinton who was elected in 2014, offered mostly tepid opposition to Virginia's climate deniers, supporting clean energy while at the same time arguing for more fracking and offshore drilling in the state. To his credit, however, McAuliffe was one of the first governors to push back after President Trump abandoned rules to cut CO_2 pollution from power plants, ordering state air regulators to propose a plan by the end of 2017 to scale back CO_2 pollution from the utility sector and increase renewable energy investments throughout the state. "The threat of climate change is real," Governor McAuliffe said when he announced the order.

In Virginia, much of the head-in-the-sand attitude toward climate change can be traced to the political power of the fossil fuel industry, especially Big Coal. Dominion Energy, the state's biggest electric power company, is also one of the biggest coal burners in America. In fact, 95 percent of the power used on the base is from Dominion, which means that US Navy operations in Norfolk are largely dependent on coal and natural gas. For the sinking base, it's a kind of fossil-fuel-assisted suicide.

The most immediate problem Norfolk faces is keeping the roads open. One study by the Virginia Institute of Marine Science identified more than five hundred miles of flood-vulnerable roads in Norfolk, Virginia Beach, and the Chesapeake region. "It's the number one problem we have," said Captain Pat Rios, who was in charge of navy facilities in the mid-Atlantic region. "If people can't get back and forth to work on the base because the roads are flooded out, we have a big problem." But the navy also has a big problem because the roads in Norfolk are not its responsibility—they are the state's responsibility. And because a large number of the men and women in the Virginia General Assembly who hold the purse strings don't believe that climate change is a big problem, they don't want to spend much money fixing it. "They find roads to fix in other parts of the state," said Joe Bouchard, the former commander of the base.

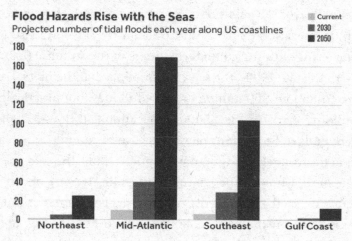

Flood Hazards Rise with the Seas
Projected number of tidal floods each year along US coastlines

Current
2030
2050

Northeast Mid-Atlantic Southeast Gulf Coast

Norfolk is one of the most vulnerable areas in the United States to repeated flooding. *(Illustration by the author)*

By far the most important pieces of infrastructure on the base itself, however, are the piers. They are the critical interface between land and sea, highly engineered concrete decks that serve as safe moorings for ships as well as access platforms for service crews. At Norfolk, most of them are as wide as a two-lane highway and about 250 feet long. In the late 1990s, navy engineers realized that the thirteen World War II–era piers at the base were reaching the end of their lifespans. In addition, because they had been built at a time when nobody gave a thought to sea-level rise, the piers were relatively low to the water. And because the sea was rising, they were getting lower every year, which made maintenance difficult. At high tide, the utilities that ran along the underside of the pier deck—electrical, steam, phone, Internet—were often immersed in water, rendering the pier unusable. To make a repair, a crew had to go out in a small boat at low

tide and bob around beneath the pier—a slow and dangerous endeavor. "It was not a nuisance problem. It was not a
minor operational issue," said Bouchard. "Sea-level rise was
interfering with the combat readiness of the Atlantic fleet."

In the late 1990s, the navy began replacing the piers. Each
one cost about $60 million—a lot of money, but practically a
rounding error in the $500-billion annual Defense Depart-
ment budget. So far, four new piers have been built, which are
higher, stronger, and better-designed than the old piers.
Bouchard, who was the commander while the first new piers
were built, said, "They were built with sea-level rise in mind."

But out on the base, nobody wants to talk directly about
spending money to deal with sea-level rise, mostly because they
are worried about drawing scrutiny from climate deniers in
Congress who are happy to redline any expenditure with the
word "climate" in it. Instead, many people in the military end up
talking about climate in much the way eighth graders talk about
sex—with code words and winks and suggestive language.

"We didn't raise the piers because of climate change,"
Captain Pat Rios told me during my visit to the base in
November. He didn't quite wink, but almost.

"Then why did you raise them?" I ask.

"Because we needed new piers. And as long as we were building new piers, it didn't cost much more to build them higher."

"But isn't sea-level rise why you built them higher?"

"That was one factor, yes. But the main thing is, we had
to build new piers anyway."

This is how conversations about climate change often go
with people in the military these days. They know better
than to talk about it directly and forcefully, lest they anger
the elected officials who fund their projects and who believe
that climate change is not a problem and that we shouldn't

be spending much time or money preparing for it, especially when there are terrorists to fight.

But building new piers, no matter how high off the water they are, is not going to save the base in Norfolk. No matter how much money the Pentagon spends on new piers and seawalls, it won't matter if people can't get to the base because the roads are underwater and nobody wants to live in the area because they're constantly flooded out and the value of their homes is spiraling down, taking with it the region's tax base, impacting everything from funding for public schools to trash pickup schedules. All of the base's critical infrastructure—water, sewage, electricity, phones—comes from off-base. "To save the base, you have to save the whole region," said Bouchard.

Climate change is not an issue the US military can afford to ignore. Drought contributed to the rising food prices that triggered the Arab Spring revolt in Egypt in 2011; it may have helped trigger the civil war in Syria. In northern Nigeria, a region destabilized by extreme cycles of drought and flooding, Boko Haram is terrorizing villages and killing thousands of Nigerians. Wildfires in the western United States are taxing the National Guard, forcing the air force to deploy planes to help out. And increasingly intense and frequent typhoons and hurricanes are pushing the military to get involved in more and more rescue and relief operations, stretching their budgets and interfering with their war-fighting capabilities.

In the decades to come, all this is likely to get worse. Regions like the Arctic, which the US military has basically ignored since the end of the Cold War, are likely to become major flash points in the territorial disputes and resource wars of the future. Within the next decade, as the ice melts away, more tourists will arrive, drilling for gas and oil will increase, new shipping routes will open. The US military will be called upon to protect American interests in a new and unfamiliar world—and one that they are poorly equipped to operate in. In the not-so-distant future, the Bering Strait—the fifty-mile-wide gap between Russia and the United States off the coast of Alaska—could become a strategic choke point in global trade like the Strait of Malacca in Asia or the Strait of Hormuz in the Persian Gulf. Rear Admiral Daniel Abel of the US Coast Guard, who is in charge of operations in Alaska, is well aware of the challenges he faces: "If you didn't know, thirteen percent of the undiscovered petroleum is up there, thirty percent of the undiscovered natural gas, over a trillion dollars of minerals are up there," he said recently. "The best way I've heard it explained is: imagine if you have the Panama Canal and Saudi Arabia's worth of energy show up at the same place in your area of responsibility. How would you embrace that?"

You can already see glimpses of a militarized future in the Arctic. In September of 2014, six Russian fighters were detected near Alaska; when US and Canadian fighters intercepted the Russian jets about fifty-five miles off the coast, still outside American airspace, the Russians turned around and headed home, but it was a close encounter—and one that happens about ten times a year. In November of 2015, a Russian sub in the Barents Sea near Greenland test-fired a Bulava intercontinental missile—the Bulava is Russia's latest and deadliest nuclear weapon. The missile has a range of about five thousand miles and can be loaded with up to ten nuclear warheads, each of which can be individually maneuvered. In this case, the test missile was pointed back toward Russia. But a Bulava launched from a sub in the Arctic could easily reach Boston, New York, or Washington, DC.

These provocations were seen by some Pentagon planners as

more than old Cold War game-playing. In their view, Putin was sending a not-very-subtle message that he thinks of the Arctic the way Americans once thought of the West—a vast, uncivilized landscape of resources that will belong to whoever stakes the first claim. In the last few years, the Russian military has built a series of new bases in the Arctic, has developed new ports in the region, is upgrading its already-impressive fleet of icebreakers (six of them nuclear-powered), and is constructing a new class of Arctic patrol vessels armed with cruise missiles. In 2007, during a year of sudden and dramatic melting of summer sea ice in the Arctic, Russian soldiers in a minisub dove ten thousand feet beneath the North Pole and planted a Russian flag in the seabed, marking it as their turf. "This isn't the fifteenth century. You can't go around the world and just plant flags" to claim territory, Canada's minister of foreign affairs, Peter MacKay, said dismissively. But in Putin's world, you can do exactly that. Especially if you have a bunch of subs with Bulava missiles.

Nobody knows what Putin's intentions are in the Arctic (including, most likely, Putin himself). Some analysts suggest that Putin's approval ratings in Russia depend on his willingness to poke the West, and that the Arctic is a good place to do it. Michael Klare, author of *Resource Wars,* a book about the coming conflicts over dwindling fossil fuel reserves, also sees the Arctic as essential to Putin's future. "He can't continue to dominate gas and oil sales in Europe without developing the Arctic," Klare argued. "The pull of geopolitics is very strong this way." Al Gore has a contrarian view. "Falling oil prices have hurt Putin domestically, and it means he doesn't have the revenues to fund a big Arctic exploration," Gore told me. "Depending on the price of oil, and how fast renewables ramp up, there is a real chance that we might not see development in the Arctic." But regardless of what happens with the oil, the strategic value of controlling—or at least projecting power in—the Arctic is unlikely to fade. As the ice retreats, there will be vast reserves of minerals to unearth (when I was in Greenland recently, it was nearly impossible to rent a helicopter—they were all booked by mining companies scouting potential mine sites), as well as shipping lanes to protect. "Its strategic value only grows," said Klare.

Whatever happens up there, it will be a big deal for the navy. "The melting ice is opening a new ocean," said Admiral Gary Roughead, who was US chief of naval operations from 2007 to 2011. "It's a once-in-a-millennium event."

Since climate change and the rising waters emerged as major risks more than thirty years ago, the fight over what to do about it has been framed mostly in economic terms. Climate deniers argue that shifting to clean energy will destroy our economy; no, climate activists argue, it will save our economy. If we limit carbon pollution and the Chinese don't, will that give them a competitive advantage? Once in a while, the moral imperative of preserving the planet for future generations is touched on. But in his 2015 State of the Union address, President Obama put climate change into an explicitly military context: "The Pentagon says that climate change poses immediate risks to our national security," Obama said. "We should act like it."

On one level, that was shrewd politics, a way of talking about climate change to people who don't care about extinction rates among reptiles or food prices in eastern Africa. But it's also a way of boxing in all the deniers in Congress who have blocked action on climate change—many of whom, it turns out, are big supporters of the military.

During the Obama administration, the Senate Armed Services Committee was made up of characters like Oklahoma senator James Inhofe, Texas senator Ted Cruz, and former Alabama senator (now attorney general) Jeff Sessions, all of whom were fierce believers in the idea that the seven billion human beings living on our planet couldn't possibly have an impact on the Earth's climate. Ditto in the House Armed Services Committee, which was now chaired by Representative Mac Thornberry of Texas, who argued in a 2011 op-ed that prayer is a better response to heat waves and drought than cutting carbon pollution.

Within the Pentagon, though, the national security implications of climate change are nothing new. In 2003, Andrew Marshall, head of the Department of Defense's Office of Net Assessment, the department's in-house think tank, commissioned a report from futurists Peter Schwartz and Doug Randall about the consequences of abrupt climate change. Marshall, who is sometimes jokingly referred to within the Pentagon as Yoda, has been a mentor to former defense secretary Donald Rumsfeld and many others. The report, titled "An Abrupt Climate Change Scenario and Its Implications for United States Security," warned that the threat to global stability posed by rapid climate change vastly eclipses that of terrorism. "Disruption and conflict will be endemic features of life," the report concluded. The physical mechanism that drives the abrupt changes described in the report—the rapid shutdown of the North Atlantic ocean circulation system—is no longer an immediate concern. But the larger point—that our national security is deeply tied to the stability of the climate—is more robust than ever. In 2014, the Pentagon's Quadrennial Defense Review, the agency's main public document describing the current doctrine of the US military, went so far as to draw a direct link between the effects of climate change—drought, rising seas, more extreme weather—and terrorism. "These effects are threat multipliers that will aggravate stressors abroad, such as poverty, environmental degradation, political instability and social tensions—conditions that can enable terrorist activity and other forms of violence," the review said.

Within the upper ranks of the Trump administration, the link between climate change and conflict is well known. In unpublished written testimony provided to the Senate Armed Services Committee after his confirmation hearing in January 2017, Secretary of Defense James Mattis said it was important for the military to consider how changes like open-water routes in the thawing Arctic and drought in global trouble spots pose challenges for troops and defense planners. He also underscored that this is a real-time issue, not some distant what-if. "Climate change is impacting stability in areas of the world where our troops are operating today," Mattis said in written answers to questions posed by Democratic members of the committee. "It is appropriate for the Combatant Commands to incorporate drivers of instability that impact the security environment in their areas into their planning."

Once upon a time, before climate change became taboo for most Republicans (and some Democrats), it was possible to have an open and straightforward discussion about the subject. Even Senator John McCain, now firmly in the denial camp, didn't hesitate to draw the connection between climate change and national security. "If the scientists are right and temperatures continue to rise," he said on the Senate floor in 2007, "we could face environmental, economic, and national security consequences far beyond our ability to imagine."

But after the rise of the Tea Party movement in 2009, which was backed by Koch Industries, the fossil fuel empire run by GOP funders David and Charles Koch, such talk vanished. Instead, Tea Party Republicans worked hard to undermine any connection between climate and national security. Case in point: in 2009, CIA director Leon Panetta quietly started the CIA Center on Climate Change and National Security. It was a straightforward attempt by the intelligence community to gather a better understanding of how climate change is reshaping the world. Among other things, the center funded a major study of the relationship between climate change and social stress under the auspices of the National Academy of Sciences, which is the most respected scientific agency in America. Some Republicans in Congress didn't like it, especially John Barrasso of Wyoming, a Big Coal state. Barrasso, who became chairman of the powerful Senate Committee on Environment and Public Works after the 2016 election, has been a tireless foe of anything that might impinge on what he sees as our God-given right to burn as much coal as we want. In 2011, he introduced legislation not only to stop the EPA from regulating carbon pollution, but also to stop the agency from even studying what is going on with the climate.

When Barrasso learned about the CIA's new climate center, he went after it. His jihad gathered momentum when Panetta was replaced as director of the agency by David Petraeus, whose central interest was in figuring out how best to use drones to kill terrorists. "We felt constant pressure to water down our conclusions," said one of the coauthors of the National Academy of Sciences report. The day the report was released, the press conference was suddenly canceled and the report was buried. A few weeks later, the Center on Climate Change and National Security disappeared.

Climate deniers in Congress have learned to go after the Pentagon where it will feel it most: in the budget. In 2014, House Republicans tagged an amendment onto the Department of Defense appropriations bill that prohibited the

Pentagon from spending any money implementing recommendations in the latest report of the UN Intergovernmental Panel on Climate Change. "The amendment had no effect on the Defense budget, since the IPCC's recommendations don't really apply to us," one Pentagon insider told me. "But the intent was clear—this is going to be war." Certainly what had been made clear was that any item in the budget that included the word "climate" was going to set off alarm bells. In 2016, the Republican-controlled House went even further, voting to bar the Department of Defense from spending money to evaluate how climate change would affect military training, combat, weapons purchases, and other needs. "When we distract our military with a radical climate change agenda, we detract from their main purpose of defending America from enemies" like Islamic State, said Representative Ken Buck, a Republican from Colorado who was one of the sponsors of the measure.

In today's political climate, open discussion of the security risks of climate change is viewed as practically treasonous. In 2014, John Kerry, a decorated war hero, called climate change "perhaps the world's most fearsome weapon of mass destruction" and likened it to terrorism, epidemics, and poverty. McCain immediately slammed him, citing the 130,000 people killed in Syria, Iranian nukes, Palestinian-Israeli negotiations: "Hello? On what planet does [Kerry] reside?" Former Republican leader Newt Gingrich, who never served in the military, tweeted, "Does Kerry really believe global warming more dangerous than north Korean and Iranian nukes? More than Russian and Chinese nukes? Really?" And he followed it up with: "Every American who cares about national security must demand Kerry's resignation. A delusional secretary of state is dangerous to our safety."

The US military, of course, is not a polar bear rescue

operation. "Their main job is to break things and kill people," said Sharon Burke, a former assistant secretary of defense and now a senior advisor at New America. God knows there are plenty of ass-kicking generals who don't care about climate change, scoffingly referring to it as "Mother Nature with a sword." But the military also prides itself on its practical-mindedness. Military leaders embraced desegregation long before the rest of the nation, in part because they wanted the best people they could find, no matter what color. "It's about the mission, not the politics," John Conger, the assistant secretary of defense in charge of military installations, told me when we talked in his Pentagon office. "It's our job to deal with the world as it is, not as we wish it could be."

In the world as it is, evidence that climate change is an engine of conflict is clear. The best example is Syria. In 2015, an exhaustive study published in the *Proceedings of the National Academy of Sciences* found that rising CO_2 pollution had made the 2007–2010 drought in Syria twice as likely to occur, and that the four-year drought had a "catalytic effect" on political unrest in the area. Herders were forced off their land, seeking food and water elsewhere. More than 1.5 million rural people were displaced, causing a massive migration into urban areas, where they bumped up against an influx of Iraqi and Palestinian refugees. When researchers asked one displaced Syrian farmer whether she thought the drought had caused the civil war, she replied, "Of course. The drought and unemployment were important in pushing people toward revolution. When the drought happened, we could handle it for two years, and then we said, 'It's enough.'"

Many military commanders don't need to read a scientific report to figure this out—they are seeing the impacts of climate change with their own eyes. Admiral Samuel

Locklear, who was in charge of all US armed forces in the Pacific, is one of the most respected men in the US military— and the one with the toughest job, with both China and North Korea to watch over. But in 2013, when a journalist asked him what he believed was the biggest long-term security threat to the region, he didn't talk about the nuclear fantasies of North Korean leader Kim Jong-un or a cyberwar with China—he talked about sea-level rise, and increasingly monstrous storms that could wipe out a small nation. The political and social upheaval we're likely to see from our rapidly warming planet, Locklear said, "is probably the most likely thing that will cripple the security environment, probably more likely than the other scenarios we all often talk about."

What made Locklear's comments so compelling was that he is a decorated war fighter, not a wonk in the Pentagon office or a globe-trotting diplomat. He wasn't reacting to a political push from the White House or repeating talking points from the secretary of defense. This was the man who was charged with the day-to-day responsibility of protecting US interests in the Pacific, which includes seven of the world's ten largest standing armies, as well as five of the seven nations that have nuclear arms.

Not surprisingly, Locklear was summoned before the Senate Armed Services Committee, where James Inhofe asked him to "explain" his remarks. And he did, calmly and forcefully, schooling the senator in how steadily increasing populations in Asia would only put more people at risk from storms and other climate-related disasters. "Okay, let me interrupt you," Inhofe said, realizing it was a losing battle. He quickly changed the subject.

What Locklear correctly foresees is that a world of climate-driven chaos is already upon us, and it's only going to get worse. Where are the limits of American power? How many failed states can we prop up, how many natural disasters can we respond to? It's one thing to plan for the invasion of Normandy Beach or the siege of Fallujah—it's quite another to plan on being the rescue squad for the entire planet. We have already spent more than a trillion dollars in Iraq and Afghanistan, with no measurable success. How much more can we afford to do? "I think we have to make some strategic choices," Admiral Gary Roughead told me. "Which parts of the world do we care about most? What are the strategic flash points? Do we want to be able to operate in the Arctic or not? What kind of world are we preparing for?"

After his tour of the Norfolk base, I joined Secretary Kerry on the flight back to Joint Base Andrews near Washington, DC. The plane, a government-issue refurbished 757 that he shared with other top Obama administration officials, was nowhere near as posh as Air Force One. In fact, Kerry's compartment had a real *Dr. Strangelove* feel to it, with a metal desk bolted to the floor and an old (presumably very secure) desk phone. As we talked, Kerry took his coat off and picked at a bowl of fresh fruit, his voice hoarse after a long day. He looked exhausted, his face more drawn than usual—talking with him, it was hard not to feel the weight of the world.

We talked for a while about what we'd seen at the base, and about his hope for the Paris climate talks, which were coming up in a couple of weeks.

I told him that no matter how things turned out in Paris, it seemed to me that America still had a long way to go before it came to grips with the scale of the threat the world faces.

"We do have a long way to go, because we still have people in the United States Senate who even deny its existence," he said bluntly. "How do you mobilize your government in a democracy when part of your democratic process is grid-locked and frozen and, in some cases, ignorant?"

That was putting it kindly. Even in his darkest night-mares, I doubt Kerry could have foreseen that, in just over a year, Rex Tillerson, the CEO of ExxonMobil, the oil giant that has spent decades denying, confusing, and underplay-ing the risks of climate change, would be sitting in his chair.

I pointed out to Kerry that, despite thirty years of climate conferences and talk about the need for clean energy, global CO_2 levels are still climbing.

"Because we're trying to turn around the largest oil tanker ever built."

"Human civilization, you mean?"

"Yeah," he said. He looked out the window at the gray clouds below. "And that is a very big challenge."

10. CLIMATE APARTHEID

NOBODY KNOWS FOR sure how many people live in Lagos, Nigeria. The United Nations' official count is 13 million, but Lagos officials say it's closer to 21 million. When you are in line at the city's prehistoric airport, it feels like 30 million. Whatever the most accurate number is, everyone agrees that Lagos is one of the fastest-growing megacities in the world, with a growth rate ten times faster than New York or LA. It is also a city that is sharply divided between rich and poor. About 70 percent of the population lives on $1.25 a day or less, while the top 2 or 3 percent live behind walls in Beverly Hills–like estates. A good percentage of those people made their money in oil. Nigeria has by far the largest oil industry in Africa, producing, on average, about two million barrels of crude a day.

Lagos is a delta city, built around a lagoon, much like Venice. Like most delta cities, it is flat and low-lying, with the majority of it built on land that is less than five feet above sea level. The city's infrastructure, such as it is, is poorly designed to deal with flooding and storm surges. Beaches are washing away, the sheet-metal seawalls in the harbor are corroding like rusty tin cans. Flash flooding in the summer of 2012 shut down the city for a week. Even a brief rain creates car-wheel-deep lakes in the streets of Victoria Island, one of the city's wealthiest neighborhoods. Flooding is also a grave public health hazard. This is a city of 13 (or 21) million people and no municipal sewage system. In the slums, kids develop rashes and pinkeye after floods, and cholera outbreaks are not uncommon.

In the midst of all this, a new city is rising along the waterfront. It's called Eko Atlantic, and when I visited in late 2016, it was still a work in progress—basically a two-square-mile platform of new land that had been built in front of Victoria Island. When it is finished (or, more accurately, *if* it is finished—the devaluation of Nigerian currency, as well as other economic factors, has put its future in doubt), Eko Atlantic will encompass more than three square miles of new land, where, developers hope, 300,000 prosperous and technologically sophisticated people will live in sleek modern condos, fully equipped with fiber-optic Internet connections, elaborate security systems, and a twenty-five-foot-high seawall protecting them from the attacking ocean. It's a shiny new appendage to a megacity slum, one that sells itself as a new vision of Lagos—the Dubai of Africa.

What's happening in Lagos is part of a larger trend of combating sea-level rise with old-fashioned engineering muscle. On coasts and in shallow bays around the world, enormous dredging machines are pumping sand and gravel out of the bottom of the sea and creating new land. You can see it in the South China Sea, where China is rapidly turning

Eko Atlantic rising on newly built land in Lagos, Nigeria. *(Photo courtesy of the author)*

coral reefs into islands to support military bases, airstrips, and port facilities. Since its independence in 1965, Singapore has expanded its size by almost a quarter, from 244 square miles to 277 square miles. Japan has reclaimed more than 100 square miles of land in Tokyo Bay alone. This is not a new idea, of course. It is essentially how Carl Fisher built Miami Beach, how Lower Manhattan expanded out into its rivers, and how the coastal tribes like the Calusa built shell middens a thousand years ago.

Thanks to all this human engineering, over the past thirty years, the Earth has gained more land than it has lost. Using satellite data from Google Earth, researchers at Deltares, a Dutch research institute, found that coastal regions have gained a net 5,237 square miles of land since 1985—an area roughly the size of Connecticut. "We have a huge engineering power," Fedor Baart, an author of the study, told me.

In China, he said, "the coastline all the way from Hong Kong to the Yellow Sea has almost been redesigned."

This statistic—the coastlines are growing, not declining—is sometimes used by climate skeptics to argue that sea-level rise is not such a big deal. If we lose some land to the sea, we can just build more. But simply to say that the total landmass on the planet is growing and not shrinking doesn't tell you where the land is increasing or what it is being used for. And it ignores the obvious fact that at some point very soon, acceleration of sea-level rise will cause low-lying land to be engulfed at a rate far faster than dredging machines can create it anew.

The day I visited Eko Atlantic, it was raining, and the streets of Lagos were flooded several inches deep with foul-smelling black water. My taxi pulled up at Eko Atlantic's sales office, which was behind a gated entry with a security guard. The office was a low, unremarkable building on the edge of Victoria Island, right where the new land began. When I stepped into the lobby, I was greeted by Yuki Omenai, a sturdy-looking Nigerian in his late thirties who was dressed in brightly colored Senator, the traditional attire of Nigeria. "Welcome to the future of Lagos," he said to me in perfect British-inflected English. Omenai, who is from a wealthy and politically connected Nigerian family, explained he has worked as a town planner at Eko Atlantic since 2010, shortly after they started selling building lots on the yet-unbuilt land.

Omenai led me into a showroom nearby, where the whole development was mapped out on an enormous table. Each lot was marked, each street, each tree. Omenai explained that Eko Atlantic will be a mix of commercial and residential buildings, that it will have its own natural gas power plant, its own water supply, its own schools, and, of course, its own security force. On the walls were artistic renderings of what the development will look like—glitzy high-rises, traffic-free streets, wide promenades where people will enjoy the fresh sea air. The showroom reminded me of condo sales offices in Miami, where a beautiful life is imagined for you, if you will just get out your checkbook and put down a deposit.

I pointed to a random plot on the map of the development. It was 6,000 square meters. "How much is that?"

"Eighteen million euros," he told me.

I pointed to a smaller one—the smallest on the map: 2,500 square meters. "How much is that?"

"Six million euros."

When I looked a little shocked, he reminded me that this land would be bought by developers who would build condo towers, not single-family homes.

"How many people do you imagine will be living here when the project is completed?" I asked.

"About three hundred thousand," he told me.

"I didn't know there were that many rich people in Lagos," I said.

"Well, there are not. These condos will be bought by middle-class people, working professionals. That is our target audience."

When I asked if investors in Eko Atlantic were concerned about sea-level rise, he said, "Oh, very much so," then guided me over to look at a schematic for the seawall—the Great Wall of Lagos, it is sometimes called—they are building to protect the development. When it's complete, it will be eight miles long, a twenty-five-foot-tall necklace of granite and concrete draped across the soft neck of Lagos.

"We are acutely aware of climate change and the dangers

of the sea," he explained. "People who live here want to be sure they are safe."

A few minutes later, Omenai and I jumped into his SUV and went for a drive around Eko Atlantic. The rain had stopped, and we drove along a frontage road, then passed through a security checkpoint—the road rose up, and then we were on the new land. It looked like a barren prairie, with a few condo towers sprouting from the dirt. The road broadened and was paved with interlocking gray bricks. The sidewalks were wide and planted with rows of young palm trees, still held up with stakes. "Our chairman is a micromanager," Omenai told me as we drove. "He made them change the sidewalk color three times. And he made them pull out some of the trees and plant different ones. He is involved in every detail. He wants to get it all just right."

"Our chairman" refers to Gilbert Chagoury, the chairman of the Chagoury Group, a collection of companies based in Nigeria that are involved in everything from real estate to trucking to bottled water production. A new subsidiary of the group, South Energyx Nigeria, was created to lead the development of Eko Atlantic.

Chagoury, who is in his late sixties, was born in Lagos but grew up in Lebanon and Great Britain. He is best known for his role as a fixer for former Nigerian dictator Sani Abacha. In a 2010 investigation by *Frontline,* Nigeria's top anticorruption prosecutor called Chagoury "a kingpin in the corruption" that defined Abacha's regime. During the 1990s, the report suggested, Chagoury had helped Abacha steal billions of dollars in elaborate bribery and corruption schemes, while pocketing some for himself too. He used his money to build the Chagoury Group, as well as to buy respectability, eventually becoming a high-profile philanthropist and a friend of

Bill Clinton (Chagoury gave millions to the Clinton Foundation). It's hard not to admire the gamble Chagoury took in building Eko Atlantic, investing billions to create new land to sell to developers. "Eko Atlantic is entirely financed with private money," Omenai told me. "There is no government money involved." But because Eko Atlantic has been designated an economic "free zone," it also won't directly contribute taxes or other benefits to the city of Lagos or the larger Nigerian economy. Even more than other gated communities, Eko Atlantic says to the world, *No, we are not all in this together.*

Omenai drove to the Great Wall, where we got out and looked around. The Gulf of Guinea, one of the most dangerous waters in the world, rife with piracy, kidnapping, and hijacking, spread out to the horizon. Omenai pointed to a man rowing a small boat along the wall below us. "See how small he looks?" he said, as if to say, "See how vulnerable he looks?" Omenai proudly touted the engineering of the wall: when it's completed, its five miles will have been constructed with 100,000 concrete blocks, each weighing five tons. An accurate scale model of the wall was built in a lab in Copenhagen and tested against the worst storms in a thousand years. "Global warming and sea-level rise were all factored into this," he said. "We really wanted to create a safe haven."

We drove over to a development called Eko Pearl Towers. The first tower to be completed was a boxy building that would make Frank Gehry weep over the sorry state of the human imagination. We parked underground, making our way through a black marble corridor to the elevator. There were security cameras everywhere.

We stepped out on the nineteenth floor and toured a model condo with a leather sofa and chairs, an LG flat-screen

TV, and hip-looking modern art prints on the walls. We walked out onto a large balcony with sweeping views of the city and the Gulf of Guinea. Below, workers were putting finishing touches on an Olympic-size swimming pool.

"We want to redefine how Lagos lives," Omenai explained. "This is the new Lagos."

I looked back at the old Lagos and thought of the millions of people who live there in shacks and cheap concrete buildings that flood with every high tide. I wondered how safe I would feel up here on a leather couch on the nineteenth floor as old Lagos drowns.

When it comes to megacities with the most at risk from sea-level rise, Lagos doesn't crack the top ten in potential economic losses. Guangzhou, Shanghai, Kolkata, Mumbai, and other Asian cities are at the top of the list. Lagos doesn't rank with these cities because, in strict economic terms, the infrastructure along the coast isn't worth much compared to a place like Shanghai.

But economic losses are only one way of thinking about the consequences of sea-level rise. The number of people who may be displaced—in other words, potential climate refugees—is another. When you factor in future population growth, Lagos is near the top of the list of places to worry about. By 2050, the city is projected to have 30 million people. How many of those will be swamped by rising seas and forced to flee? Various studies have come up with numbers ranging from three million to eight million. Whatever the number, you only have to spend a few hours in Lagos to understand that sea-level rise will displace a lot of people, and those people are going to have to go somewhere.

Like other places in the world, sea-level rise is already

hitting Africa hard. West Africa is particularly vulnerable, especially the four-thousand-mile-long sub-Saharan coastline that stretches from Mauritania down to Cameroon. It's mostly low-lying and sandy—in some places the sea is eating away more than a hundred feet of land in a year. In a region where 30 percent of the population lives along the coastline, according to the World Bank, this is a potentially catastrophic problem. "In West Africa, infrastructure and economic activities are centered along the coastal region, so as sea levels continue to rise, it threatens our very existence and source of income," said Kwasi Appeaning Addo, a professor in the University of Ghana's Department of Marine and Fisheries Sciences. "We are sitting on a time bomb."

Lagos is not the only city at risk. In Accra, the capital of Ghana, low-lying areas of the city now flood every year during the rainy season. Parts of Nouakchott, the capital of Mauritania, already have lost up to eighty feet of beach every year, and erosion has already damaged several hotels in Gambia and Senegal, as well as an important water treatment facility in Cotonou, Benin's economic hub. On the outskirts of Lomé, the capital of Togo, rows of destroyed buildings line the beaches.

And it's not just homes and businesses that are being swept away. Western Africa is also suffering from the same problems related to rising seas as places like South Florida and the Marshall Islands, such as salinization of the soil and contamination of drinking water supplies. In Ghana, breeding grounds for sea turtles are disappearing. UNESCO-protected colonial forts along the coasts of Ghana and Ivory Coast that served as conduits for the slave trade are at risk of being swallowed by the sea. As the coast dissolves, the shoreline often becomes steeper, making survival more and more difficult for villagers who rely on fishing from small boats near the shore for their livelihood. Now they need bigger boats to go farther into the ocean—a necessity but unaffordable for many. Instead, some villagers turn to sand mining—gathering sand to sell for construction of concrete blocks and other building materials—which is illegal. "Some of our children go mining as soon as they come back from school, in order to gain some money," one resident of Agbavi, a small community in Togo, said recently. "People are hungry, and small kids are forced to steal. We are suffering a lot."

I didn't go to Nigeria to see the coast of Africa being washed away. I went to Nigeria because I thought I'd found an elegant solution to sea-level rise in Lagos's water slums. While I was reporting this book, I'd seen a picture of a floating school that Nigerian-born architect Kunlé Adeyemi had designed in one of Lagos's water slums in 2013. It was a simple, elegant structure, one that suggested we could solve the problem of living with water if we just thought about it a little differently. The school looked like a floating triangle. The base was made of 250 blue barrels lashed together; the structure itself was wooden, with a metal roof and open walls. The bottom floor was big enough for community meetings; the second floor held two classrooms. It couldn't have been simpler. Yet because it was so elegant in its simplicity, and so hopeful in its aspirations, it attracted worldwide media attention, won numerous architectural awards, and made Adeyemi into a star. The *Guardian* called it "a beacon of hope." Intrigued, I arranged a trip to Lagos to see it for myself.

Before I left, I drove up to Ithaca, New York, where Adeyemi was teaching a class on the problem of affordable housing in Lagos. I sat in on his session, where students were showing off

The Makoko floating school. *(Photo courtesy of Kunlé Adeyemi/NLE 2013)*

designs for simple structures, many of them variations of pole houses that could be built above water. Adeyemi sat at the front of the class, listening, provoking students with a few questions. He is a soft-spoken guy, in his forties, handsome, quiet, dressed in a white shirt and jeans, his head shaved.

After class, Adeyemi told me that the idea for the floating school came to him in 2011, when he began thinking about solutions for affordable housing. "I started looking at vernacular architecture, how do everyday people build cheap," he recalled. Adeyemi grew up in Kaduna, a city in northern Nigeria, where his father was a successful architect, designing houses, public buildings, and hospitals in the area. When he was eighteen, Adeyemi moved to Lagos and attended the University of Lagos, where he studied architecture. In 2002, he landed a job at OMA in Rotterdam, the firm founded by Rem Koolhaas. He traveled to the United States to study at

Princeton University for a few years, then returned to the Netherlands, where he started his own firm in 2010. He called it NLÉ, which is Yoruba for "home."

"My romance with water and cities began in Amsterdam," Adeyemi recalled. "To live in a city where there is so much water, to see it every day, made me think differently about a lot of things." While he lived in Lagos, he had heard about water slums—he had even seen them from a distance every time he crossed over the city on the Third Mainland Bridge— but he never spent any time in one until he began his research on affordable housing in 2010. When he visited Makoko, the largest of Lagos's water slums, he saw a whole world on water: schools, churches, machine shops, and tens of thousands of people living in shacks on stilts. "It was shocking to see people living like that," he told me. "It was also inspiring. They were doing so much with so little."

While he was in Makoko, he learned that because the school was built on reclaimed land, not elevated like other structures, it flooded a few times a year. "I asked them if I could help and build something new," he recalls. Adeyemi's initial idea was to build the school on stilts, like most of the other buildings in Makoko. But then in July 2012, just as he was about to draw up plans for the school, a combination of high tides and big rains hit Lagos. "The entire city was completely flooded," Adeyemi recalled. "There was water everywhere. It just occurred to me, this is happening here, water is everyday reality. Here I am, trying to solve the problem of flooding, and it occurred to me—make it float! That way, it doesn't matter how high the water is, it can move with it. That's when I realized I wasn't just building a school, but beginning to address an issue that I have now become obsessed with—climate change."

Floating structures are nothing new (they're also known as boats). But as seas rise, architects and city planners are thinking differently about their usefulness and design. At nearly every sea-level rise conference I've attended, there have been architectural musings about living on the water. Some people are experimenting on their own. In Mexico, a man named Richart Sowa has made a floating island out of 250,000 used plastic bottles stuffed into recycled fruit sacks. He planted mangroves and palm trees on it, built a two-story house out of wood and fabric, and calls his plastic bottle island an eco-paradise. Then there's the Seasteading Institute, which imagines an entire city at sea, far from the hands of government. The institute was cofounded by Peter Thiel, the eccentric billionaire who sits on the board of Facebook and campaigned for Donald Trump during the 2016 election. For seasteaders, offshore settlements are a kind of libertarian dream, a new city-state where the old rules don't apply. The institute recently spun off a for-profit company called Blue Frontiers, which hopes to build a laboratory and living spaces on a series of fourteen floating platforms in a lagoon in Tahiti. The proposed project, which the company sees as a prototype for more ambitious settlements on the open sea, includes floating solar panels, a high-speed internet connection, and its own cryptocurrency, called SeaCoins.

For Adeyemi, the Makoko floating school was a first sketch of a new way of living: "You can scale it up, scale it down, beginning to create solutions that exist on water, across water, in water. We need to learn to live with water, not fight against it."

Unfortunately, I never got to see the floating school. Not long before I arrived in Lagos, a big storm hit the city and the school collapsed (fortunately, no one was in it at the time). The structure, it turned out, wasn't so well-built after all, and had been poorly maintained. Adeyemi told me later that the building was a prototype and was never intended to last long. Still, the collapse was a major embarrassment for Adeyemi, who had won an award for his floating school at the Venice Architecture Biennale just a few weeks earlier. The *Guardian* described the collapse as "a serious blow to the future of the remarkable floating city."

I rode into Makoko in a keke, a three-wheeled vehicle that looks like a golf cart (only about half of the slum is permanently on water—the rest is land-based). Squeezed in the backseat with me was Fred Patrick, a muscular, neatly dressed man who grew up in a nearby slum that was recently demolished by the government in an attempt to "clean up" Lagos. Patrick now attended law school and worked as an organizer at Justice & Empowerment Initiatives, a non-profit, non-governmental organization that works with poor communities on urban development and provides free legal aid and education.

Entering Makoko was a descent into what appeared to be total chaos, but the place was better understood simply as a world that obeys rules that are hidden from a Western visitor like me. The streets were crowded with vendors selling rice, grilled corn, wallets, belts, shoes, newspapers. Cars and trucks inched along in every direction, motorcycles darted in and out. Women in brightly colored dresses balanced baskets of melons on their heads. Kids hawked DVDs to passengers in taxis stuck in traffic. Diesel fumes from buses and trucks and nearby generators filled the air. About 300,000 people live in the city's water slums, jammed into a shamble of shanties and decrepit concrete-block buildings, often with an extended family of ten living in a room the size of a closet in an Eko Atlantic condo. But despite this—or maybe because of this—everyone seemed to get along, and I saw more evidence of tolerance and patience in Makoko than in perhaps any place I have visited.

It was a sunny day, and it hadn't rained for a week. Still, water was everywhere, pooling on the dirt streets in wheel-deep sinkholes. We zigzagged through the water, then stopped on the banks of a man-made canal. A sour, chemical smell rose up from the water, which was littered with thousands of plastic bottles and bags. I remembered the advice of one friend who had visited Makoko: "One thing you don't want to do," he told me, "is fall into that water." When I looked down into the canal, I saw a dead piglet floating by.

We rented a boat for 300 naira—less than a dollar—and paddled away. Within minutes, we were deep in an elevated city, a community on stilts. Some of the houses were shacks, with walls of burlap and driftwood, while others were slum mansions, brightly painted, with two stories and rooms added on. The canals were crowded with boat traffic—kids hot-rodding around with their friends, women paddling boats full of rice and vegetables. We passed a machine shop, where men gathered, shirtless, working on an engine; a mill, where corn was being ground into mash; a small blue hut with a sign that said HAIR SALON. We motored past a school where kids sat at desks, thirty feet above the water, and churches that were built on elevated piles of sand. Kids yelled at us as we floated by; others stared at me ("They have never seen a white person before," Patrick explained). We saw people napping, washing clothes, repairing fishing nets. In short, living life on the water.

After about a half hour of paddling, we pulled up at the side of one of the better-kept houses in the lagoon. The walls were made of thin strips of bamboo, and the roof of palm

Life in Makoko slums is well adapted to water. *(Photo courtesy of the author)*

fronds. We climbed out of the boat onto a small front porch, where a man named Gerard Avlessi greeted us. He was in his early fifties and was dressed in dark brown traditional Nigerian clothes. Nearby, a dozen or so people sat in a circle, talking—Avlessi nodded to them and explained that these were his family and apprentices. "He is the village tailor," Patrick explained to me.

Avlessi invited me in and I took a seat on the couch. On the far wall, in the most visible place in the house, was a series of pictures of Christ and the Virgin Mary; a manger scene made out of fabric and doll-like figures hung in the corner. A thin red carpet covered the floor. Avlessi sat on the other side of the couch and his two-year-old son crawled up into his lap, naked. Avlessi's fifteen-year-old daughter joined us, wearing a beautiful green dress. Through the spaces between the bamboo walls, I could see ripples of black water.

I asked Avlessi how long he had lived here.

"Twelve years," he said.

I complimented him on the house. The room we were in was large, maybe twelve feet by twelve feet, with a seven-foot ceiling. There were other rooms above, and a workshop off to the side. He told me that nineteen people lived in this house, including his wife and kids and apprentices. "Sometimes," he said, "fifty people live here."

That was hard to imagine: the whole place, including his workshop and the porch, was not much bigger than a fort I built in my backyard as a kid.

"We are very comfortable here," he explained.

"Did you build the house yourself?" I asked.

He smiled and said, "Yes, with some help from my family."

"How long does it take to build a house like this?"

"If the materials are available, it takes about a week."

Using hand motions, he pantomimed using a hammer to drive the poles into the sandy bottom of the lagoon. The poles are driven about nine feet into the sand, and he said they last about fifteen years before they rot and need to be replaced.

"How high is your house above water?"

"About four feet," he said.

"Do you have any trouble with flooding?"

He shook his head. "No trouble with water," he said.

"Storms?"

He shook his head. "No problem."

Patrick pointed out that if a house was getting water in it, it was easy to just raise it higher. "It is very simple to do," he explained. "We can do it in a few days. We do it all the time."

I thought about my visit to Eko Atlantic a few days earlier.

The conventional wisdom says climate change will hit the poor harder than the rich. And in many ways, that is true: the rich live in better houses, have access to better health care, and have the money to leave if the going gets tough. But they are also helpless if their phone battery dies. They can't change a tire, much less build a house on water or raise it in a few days. Technology gives us power, but it also enfeebles us. Listening to Patrick and Avlessi, I thought, "These guys know how to survive."

But of course there are many threats that people in Lagos face beyond sea-level rise. And for these, the residents of Makoko are not so well adapted.

This became clear when I asked Avlessi, "Are you worried about climate change and sea-level rise and how it will affect your life here?"

He shrugged. "I am not afraid of water," he said, then paused. "I am afraid of our government."

I understood why. A few weeks before I arrived, Lagos's governor, Akinwunmi Ambode, had issued an order that people in all waterfront slums would be evicted within seven days and their homes would be bulldozed. The government argued that they were clearing these people out because they were kidnappers and thieves and good-for-nothings. While the notion of giving people better places to live is admirable, the government of Lagos does nothing to help people in these communities start a new life after they have been evicted. "The police simply arrive and tell people they have two hours to get out," Megan Chapman, the cofounder of Justice & Empowerment Initiatives, told me. Then the chain saws are unsheathed, and within hours, tens of thousands of people are homeless. Patrick had told me that a few months earlier he received a message when he was in school that the authorities were coming to bulldoze the slum he had grown up in. "By the time I jumped on a bus and made it back there, my family home was gone," he told me.

Avlessi looked grave when I asked him how he dealt with the threat of eviction. I looked at the furniture, images of Jesus on the wall, the white lace tablecloth one of his daughters had laid out on a table before she served me a Coke. Makoko might be a black-water slum, but it is also a blueprint for how to live in the age of rapidly rising seas. In a rational world, the city of Lagos or the government of Nigeria or some wealthy oil baron would see this, would invest a few hundred thousand dollars in improving sanitation for the people in Makoko and hold them up as model citizens of the future. Instead, their houses will be chain-sawed or burned and they will be forced to live on the streets or jam themselves into tiny rooms in shabby concrete-block buildings, which, like virtually all buildings in Lagos, have been built at sea level and are therefore doomed in the coming years, creating a new generation of refugees who may or may not turn to crime or terrorism, but who will pay for the stupidity and greed of others with the health of their children and their own brutally short lives.

"We are all worried about this," Avlessi told me. "But there is nothing we can do. In Lagos, after God, there is government." He bounced his young son on his knee and looked off into the distance. "If it were possible to take a boat to God, and report Lagos state government to God, I would have done that."

Three weeks after I left Nigeria, police entered a nearby slum and burned it to the ground, leaving 30,000 people — mostly families with young children — homeless. A few months later, thousands more were displaced when police stormed another community, Otodo Gbame, firing bullets and tear gas and forcing residents to flee on boats. A twenty-year-old man named Daniel Aya was shot in the neck while he was trying to rescue family belongings and later died. The homes were all burned to the ground.

11. MIAMI IS DROWNING

Sᴜɴsᴇᴛ Hᴀʀʙᴏᴜʀ ɪs a neighborhood on the bay side of Miami Beach, which is the low side of the already low barrier island. A hundred years ago, it was a mangrove swamp. Today, million-dollar houses and condos have a romantic view to the west, over Biscayne Bay and toward the ever-changing towers in downtown Miami. Until recently, it was a not-so-hip area, with cheap 1950s apartments, muffler shops, and the City of Miami Beach's only public boat ramp. A couple of big condo towers went up in the 1980s, but the neighborhood remained mostly forgotten until it was discovered by Scott Robins, a well-known Miami Beach developer, as well as his friend, Philip Levine, who had made a fortune selling specially produced magazine and TV shows to cruise ships and, like everyone else who made a small fortune in Miami, was trying to turn it into a larger fortune by dabbling in real estate development.

Robins and Levine brought in restaurants, coffee shops,

and a new Publix supermarket. Old apartment buildings were torn down, new condos went up, and real estate prices boomed. But there was one problem: Sunset Harbour, more than most other places in the city, was prone to flooding during high tides or big rain events. Because it was on the low side of the island, high water tended to arrive there first and hang around for a while. In 2012, I visited Sunset Harbour during the annual king tides, which hit South Florida every year around mid-October. King tides are driven by a particular alignment of the sun and moon and Earth that maximizes the gravitational tug on the oceans, as well as changes in the Gulf Stream current and the way the heat of a long summer causes the ocean to expand. That October, I waded through water up to my knees in the streets of Sunset Harbour. Residents were enraged. The shop owners I talked to were on the verge of moving out. In fact, it was trudging

Enjoying the high waters in Miami Beach. *(Photo courtesy of Maxtrz/Creative Commons)*

through the water in Sunset Harbour that made me aware for the first time of the real-time risks Miami faced from sea-level rise.

Not long after that visit, several things happened. The first was that cruise-ship-entrepreneur-turned-developer Philip Levine decided to run for mayor. He was a slick guy in his late forties, handsome, personally reserved but politically outgoing, and not afraid to knock on every door in Miami Beach and ask for a vote. He made flooding a central campaign issue, even managing to exploit it for laughs in a TV ad that showed Levine and his boxer, Earl, both of them outfitted with life jackets, kayaking through the streets of Miami Beach. Levine won the 2013 election, and suddenly, after much denial and bureaucratic foot-shuffling, the city began to take action on sea-level rise.

As it happened, the city had hired a new chief engineer named Bruce Mowry a few months before Levine took office. Mowry, a hard-charging Southerner who was sixty-three years old when he took the job, prided himself on his ability to get things done. At the urging of the mayor, he took a hard look at the city's storm water master plan, which had been produced a few years earlier with the help of AECOM, a global engineering and consulting firm. In Mowry's view, the plan understated the risks the city faced from higher tides and rising seas, but he got to work implementing the parts of the plan that would have the biggest impact first. In many areas of the city, flooding was caused by seawater backing up though the wastewater drainage pipes and bubbling up in the streets through manholes and sewer grates. To fix that, Mowry began installing one-way check valves on the drainage pipes that would stop the water from backflowing into the streets, as well as installing big pumps in low-lying areas of the city to drain the water that did accumulate. To finance all this, the city commissioners pushed through $100 million in bonds by raising the storm water fees on residents' utility bills by about $7 (this was only a down payment—Mowry says the final price tag for the city's flood abatement plans will likely rise to $500 million). In addition, Mayor Levine did what politicians always do when they are confronted with a politically complex issue—he formed a blue-ribbon panel and appointed a pal to oversee it. And who better for the job than Scott Robins?

Mowry and Levine spent most of 2014 working furiously to show progress on flooding in Miami Beach. Their goal was to show significant headway by the time the king tides hit the following October. Levine understood that because he had made sea-level rise such an issue in his campaign, his political future depended on his ability to deal with it. The media had caught on too. In 2013, I wrote a long story for *Rolling Stone* that highlighted the risks the city faced, and in the coming months, the *Washington Post,* the *Guardian,* and many other publications followed suit.

When the king tides arrived in 2014, it was clear that the mayor's political gamble had paid off. The sea rose, but the check valves worked and the pumps sucked up the water. Sunset Harbour was mostly dry (it helped that the king tides were unexpectedly low that year, because of factors not related to sea-level rise). The mayor had invited a collection of dignitaries to a little park in Sunset Harbour to celebrate his accomplishments, including EPA administrator Gina McCarthy, Florida senator Bill Nelson, and Rhode Island senator Sheldon Whitehouse. They all praised the city's efforts and used the occasion to call for deeper cuts in carbon emissions. In the world of climate change, it was one of those rare uplifting events. *Yes, Miami Beach is in danger—but look what a little sweat and ingenuity can do!*

A few days later, I went to see Robins at his office in South Beach. He was friendly and chatty, with the air of a Miami hipster who has recently become serious about civic matters. We talked for an hour or so about the challenges Miami Beach faced and how the city might deal with it in the future.

Two things were memorable about that encounter. The first was that, despite the fact that Robins was head of the mayor's Blue Ribbon Panel on Flooding Mitigation, there was no danger that anyone would mistake him for a climate scientist. Case in point: when we talked about future rates of sea-level rise, he pulled out a tide chart from that week and remarked on the difference between the levels of the projected tides and the actual tides. In some cases they were off—higher or lower—by several inches. Robins asked, "If we can't even predict the tides this week, how are we going to predict sea-level rise in the future?" This was, I realized, a twist on a familiar point of confusion (as well as a talking point for climate deniers), which is that if we can't predict the weather next week, how can we predict the climate in twenty years? I pointed out to him that tides and sea level were very different: the daily tides were based on small and chaotic changes of winds and currents, while sea-level rise was about long-term averages. But Robins didn't seem interested in such nuance. He just said, "Yeah, I understand," and then changed the subject.

The second thing I learned was that Robins had a very clear plan for how Miami Beach was going to deal with sea-level rise, real or not.

"We're going to raise the city two feet," he told me, point-blank.

I was startled to hear this put so bluntly and so decisively. "What do you mean, you're going to raise the city two feet?"

"I mean we're going to raise all the streets and buildings in Miami Beach two feet. It might take some time, not going to happen overnight, but that is what we are going to do."

"Do you have an engineering plan for this?"

"Not yet, but we're working on it."

"Do you have any idea what it will cost?"

"No, but we will figure out a way. There is a lot of money here."

I pressed him for more details, but he offered none. So I was more than a little surprised a few months later when I heard Mayor Levine announce that the city planned to raise a few streets in Miami Beach. And the place they were going to start, naturally enough, was Sunset Harbour.

In the 1850s, Chicago was a booming outpost on the prairie, a city growing so fast that nobody gave much thought to prosaic things like urban planning or infrastructure. Wooden buildings gave way to five-story brick buildings, including big, fancy hotels. But as the population grew from 20,000 to more than 100,000 in just a decade, it became clear that chaotic build-whatever-you-want development couldn't go on. The city's buildings had been designed at various heights, so the wooden sidewalks were a jumble of stairs and planks, up and down and across mud holes that were, as a popular saying of the time goes, deep enough to drown a horse. Even worse, the city was built on a bog too near the shore of Lake Michigan — when it rained, it flooded. And because the newly built city had no municipal sewage system, those floodwaters were often contaminated. In 1854, a cholera epidemic killed 1,424 Chicagoans; between 1854 and 1860, dysentery killed 1,600.

To solve the flooding and sewage problems, city engineers came up with an innovative solution. Instead of digging in the muck to lay sewage pipes underground, they laid the sewage pipes on top of the ground and covered them by raising the entire city about eight feet. Ambitious engineers like George Pullman, who would later make his fortune with the invention of the Pullman railroad car, were soon using thousands of wooden corkscrew jacks to elevate five-story brick buildings without even cracking a pane of glass. Pullman's most famous feat was elevating the Tremont House, the finest hotel in Chicago, which took up almost an entire city block, while the guests remained in the hotel. All in all, the raising of Chicago took about a decade, and it was a triumph of urban engineering. The contaminated bogs were eliminated, public health improved, and Chicago became one of the fastest-growing cities in the world.

But Miami Beach circa 2017 is not Chicago circa 1860. For one thing, Chicago was still a new city, with dirt roads and little infrastructure. Miami Beach has water lines, wires,

Raising the Briggs House hotel with corkscrews in 1857. (*Photo courtesy of Wikimedia*)

sewage systems, paved roads, concrete sidewalks. Raising (or moving) a structure in old Chicago was relatively cheap compared to the cost of building anew. In Miami, in most cases, it's the opposite.

In Miami Beach, Mayor Levine and Mowry decided to get started anyway. Their strategy was to raise one street at a time, and to raise only the street and sidewalks. Then, over time, they hoped that owners of the buildings on the street would raise or rebuild the structures at a new, higher level.

By the end of 2016, about twenty blocks had been elevated, mostly in and around Sunset Harbour. The effect was surreal. The restaurants and shops were down at the old levels, but the streets and sidewalks two feet higher, so that to get to a restaurant, you had to step down into a sort of sunken patio. To get to the supermarket, which was originally built at a higher level, you had to step up. There were odd bumps in the road where the old elevations met the new. Beneath it all was a newly engineered pump and drainage system that was supposed to keep the low areas dry during big rains or king tides.

That was the idea, anyway. A few weeks before the king tides in 2016, Hurricane Matthew rolled up the East Coast. The storm didn't target Miami directly, but the city was hit by torrential rains. As it happened, several of the new pumps in Sunset Harbour were offline. The neighborhood flooded with several feet of water, just like old times.

A few weeks later, during the king tides, there was water in the newly elevated neighborhood again—this time, the pumps were all working, but the king tides, combined with another big burst of rain, overwhelmed the system. One night, in the rain, I drove around Miami Beach and saw water not just in Sunset Harbour, but everywhere. It was car-wheel-deep

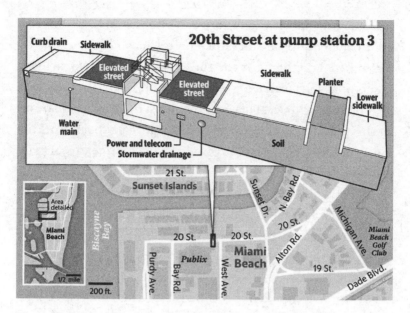

Diagram of elevated street and pump station in Sunset Harbour neighborhood of Miami Beach. *(Illustration courtesy of City of Miami Beach)*

in front of the Fontainebleau Hotel and around the Lincoln Road pedestrian mall (which also happened to be the priciest commercial real estate in the city) and lapping up against temporary flood barriers at a Florida Power & Light substation on Fortieth Street. The high water only lasted a few hours; it receded as the tides diminished. Although the reemergence of dry ground was reassuring, the flooding was terrifying to anyone who was paying close attention. It was nature's preview of the disaster film to come.

At nine the next morning, just as the next cycle of king tides was about to hit, Henry Briceño, a geologist at Florida International University, picked me up in his well-worn silver Honda Civic. The trunk was full of ice chests and plastic bottles and scientific gear. Briceño, then seventy, who was born in Venezuela and worked there as a scientist until he was run out of the country by dictator Hugo Chávez, looked like he was on his way home from the hardware store, wearing a green polo shirt with a big water stain on it, khaki pants, and beat-up hiking shoes. He was a specialist in water quality, and by the time he picked me up, he had already been working for several hours, driving around to check on teams of grad students taking water samples across Miami Beach.

Shortly after we drove off, Briceño stopped at a traffic light and looked down at my shoes, which were low-cut sneakers. "I meant to remind you to bring along some rubber boots," he said apologetically. "You won't be wanting to get your feet wet."

"I know—should have thought of that myself," I said, feeling a little foolish.

"As long as you don't have any cuts or open sores on your feet, it should be okay," he said, although that did not exactly sound reassuring.

But I knew what he meant. He was going to be wading through waters that were likely to be polluted by human fecal bacteria, and if I wanted to come along, I should understand the risks.

And I did. Briceño's work was important because it suggested a simple but often unacknowledged truth about sea-level rise: it ain't gonna be pretty. In urban areas, the floods that pour into the city are not going to be luminous blue waters you'll want to frolic in on your Jet Ski. They are going to be dark, smelly, and contaminated by organic and inorganic compounds, including, in some places, viruses and human shit.

Briceño's interest in Miami Beach's water quality began in 2013, when he saw neighbors wading through knee-deep water in the streets and thought, "Is that water safe?" He got out his instruments, did some testing, and found the water was definitely polluted. Not only were people walking through it, but it was also being pumped out into the fragile waters of Biscayne Bay. To better understand what was going on, Briceño organized a joint research project with scientists from FIU, the National Oceanic and Atmospheric Administration, the University of Miami, and Nova Southeastern University. They sampled the water quality at a number of discharge sites over a period of four days during the king tides in 2014 and 2015. At four sites tested in 2015, every spot had fecal levels above state limits. Along Indian Creek Drive, a main thoroughfare in the city, levels were 622 times higher than limits allowed by the state. A storm drain outfall at Fourteenth Street measured 630 times higher than allowed.

When Briceño's group presented their findings to the city in a report in 2015, city officials ignored it for nearly a year—until the *Miami Herald* got wind of it and published a story about Miami Beach's polluted floodwaters in January 2016. Many powerful players in the Miami Beach business and political world were outraged that the *Herald* had published a story that could sully the city's reputation as a vacation paradise. Mayor Levine accused the paper of running the story "in order to sell ads." During a speech at a luncheon I attended celebrating the 100th anniversary of Miami Beach, Scott Robins said, "there is someone around here saying that we are dumping polluted water into the bay. That is not true…that is a lie…he is a liar." (Later, I asked Briceño, who was in the room when Robins made his remarks, how he reacted: "I simply laughed quite loudly and said, 'that son-of-a-bitch just called me a liar…!' Then I stood up and left the venue.") Beach commissioner Michael Grieco called his report a "hit job." Mayor Levine went after Briceño personally at a city council meeting, suggesting that Briceño was trying to strong-arm the city into paying over $600,000 for a contract to test water. The Miami Beach city attorney wrote a letter to the *Herald,* demanding that they retract the article and charging that the article "recklessly and incorrectly depicts vile, unsafe water conditions around the city." The *Herald* declined to retract the story.

Briceño, whose work is widely respected by his scientific colleagues, refused to be intimidated (this is a guy whose email signature includes a quote from physicist and science celebrity Neil deGrasse Tyson: "The good thing about science is that it's true whether or not you believe in it"). "I don't know what's behind this mayor—I'm not attacking him personally," Briceño told me as we drove through the city that morning. "I just want people to know that they shouldn't be here swimming, because they are exposed. And he has been hiding this information. They are finally informing people. But they have known about it for a long time, and they have said nothing about it. Now, because of all the media attention these studies have received, they are being forced to acknowledge it."

At about that time, we arrived at Briceño's first sampling site, which was a few blocks from Sunset Harbour. We walked over to the seawall, where a couple of graduate students were dipping a plastic bottle into water that was rushing out of the outfall pipe. The water had a foul smell caused by hydrogen sulfide, a gas given off by decomposing organic material. Standing on the seawall, Briceño talked about how storm water—whether it's from high tides or rainstorms or both—cycles through the system here, washing through the streets, seeping through the ground, and ending up in the storm water drainage pipes. Although the city has a municipal sewer system, many of the pipes—especially the smaller ones that connect homes and offices to the main sewer lines—are corroded or cracked. The sewage leaks out, mixes with the floodwater, and ends up either sitting in the streets or being pumped out into the bay. "But at the end of the day, what's happening is not very complicated," Briceño said. "There is a guy up there"—he pointed to some nearby apartments—"who is shitting, and it is coming out here."

Briceño understands very well why some people don't want to hear about any of this. "The entire tourist economy depends on the quality of water here," he explained. "Tourists come here because they can swim and boat and Jet Ski. If the entire bay is smelling like this right now, I don't think you're going to want to go swim in there. So that's what it is. The economy needs to be somehow protected. I understand that. But I also understand that because of sea-level rise, everyone here, sooner or later, is going to have to move away. In the meantime, we have to have the best standard of living we can, and best quality of water, for as long as we can. If we destroy that, nobody is going to come here, and we won't have any money to raise the streets or build seawalls or do anything. So we have to keep our waters as clean as we can for as long as we can, and then use that time to plan how we are going to get the hell out of here."

Next stop on our polluted water tour: Shorecrest, a poor neighborhood in Miami, just over the bridge from Miami Beach. We arrived just as the morning high tide was reaching its peak—the water in the street was three feet deep, reaching up into people's yards, over their lawns, right up to their front doors. There were no traffic cones set up, no cops diverting traffic, no public health officials warning people to stay out of the water.

Briceño parked and began assembling his water sampling equipment and putting on his knee-high rubber boots. He noticed a guy with a camera about to stroll into the water. "I wouldn't go in that water if I were you," Briceño shouted. The guy nodded but walked into it anyway and started snapping photos.

While Briceño collected his water samples, I hopped and skipped over dry ground to a nearby apartment complex, where I found a woman named Maria Toubes staring at the incoming water from her second-floor doorstep. She was sixty-five and disabled, a hard life etched in her face. Inside, she had an eight-year-old niece whom she wouldn't let out of the house because of the high waters. Toubes explained that she lived on a fixed income and had moved into this neighborhood a few months earlier because it allowed her to save $200 a month on rent.

As we talked, the water continued to rise, pushing up the street in front of her house and into her driveway. It felt like we were about to float away.

"Have you seen the water this high before?"

"Sometimes it comes up even higher," she said.

"What do you do?"

She looked at me as if I had asked a very stupid question. "Stay inside," she said.

"Has anyone from the city or state ever been here to warn you that the water may be polluted?"

She shook her head. "Nobody has been here to tell me anything."

By the time we finished talking, the tide had risen even higher, and I was forced to wade through the floodwaters to get back to Briceño's car. When I got there, I took off my shoes and socks, rinsed my feet with bottled water, then spent the rest of the morning barefoot. I watched Briceño process his water samples, filling up a syringe with water, then pushing the water through a filter, which catches all the impurities. He stored the water and the filter in a plastic cooler to preserve the bacteria (and anything else) until he got back to the lab.

A few weeks later, he emailed me results from the water samples at Shorecrest and around Miami Beach. The indicator that the EPA uses for fecal matter in the water is *Enterococcus,* which is a bacteria that is easily and reliably traceable. The EPA standard for acceptable contamination in water is 35 colony-forming units per 100 milliliters of water. According to Briceño's tests, the floodwater in Shorecrest had 30,000 CFUs. Results for most of the sites he had sampled around Miami Beach were similarly high. To put it crudely, the water I was wading through, and that Maria Toubes and many others were living in, was full of shit.

In Miami-Dade County, the majority of homes and businesses are hooked up to a municipal sewer system, which collects waste through a network of underground pipes and sends it to a central sewage plant, where it is treated, then dumped into the ocean. In 2014, after more than 150 spills dumped more than 50 million gallons of sewage into Biscayne Bay, the county reached an agreement with the EPA to spend $1.6 billion on repairs; the county is now looking at injecting wastewater into a deep well 3,000 feet belowground. A larger problem, however, is that 20 percent of the people in Miami-Dade depend on old-fashioned backyard septic tanks. There are about 86,000 in-ground systems in Miami-Dade County, and a total of over 2 million in the state. And most of them are old and poorly maintained.

Septic tanks are basically just well-engineered holes in the ground. When you flush the toilet, wastewater drains into a concrete tank. Fecal matter and other wastes stay in the tank, decomposing into a sludge, while liquids run off into a leach field that surrounds the tank. When it's working properly, the soil around the leach field acts as a filter, removing bacteria and pathogens. Septic tanks require regular maintenance: the sludge needs to be pumped out of the tanks, the leach fields checked to be sure they are not blocked or collapsed. But the State of Florida, like most state governments, requires no regular performance inspections of septic tanks. According to the Florida Department of Health, just 1 percent of the state's systems are inspected and serviced each year. By one estimate, more than 40 percent of the septic tanks in the state aren't functioning properly.

As sea levels rise and more and more neighborhoods are inundated, those problems are only going to get worse. "As the groundwater rises, the systems don't drain properly," Virginia Walsh, a senior geologist at the Miami-Dade Water and Sewer Department, explained to me. "Leach fields don't filter out bacteria. A flooded septic system is worthless." For Walsh and others concerned about water quality and public health in Miami as sea levels rise, septic tanks are a big problem. "We have seen it before in flooded areas," Walsh told me. "The septic tanks pop right out of the ground. They begin to float."

Brittinie Nesenman, the co-owner of Jason's Septic Inc., one of the largest septic tank installers and service providers in Miami-Dade County, told me that leach fields need to be a foot above the water table to function properly; otherwise, the leach fields can collapse. "We are definitely getting more calls for repairs because water tables are rising," Nesenman told me.

Erin Lipp, a microbiologist at the University of Georgia's

College of Public Health, participated in an experiment with septic tanks in the Florida Keys, where the water table is high. Residents in the Keys had noticed coral dying off, and increasing algae blooms. When researchers tested the waters in several canals, they found evidence of sewage pollution. It wasn't hard to guess where it was coming from. Lipp and her colleagues put viral tracers in a toilet, flushed, and eleven hours later, the tracers showed up in a nearby canal. "And this in a place where everyone thought their septic tanks were working fine," Lipp explained.

High levels of nutrients in leaking sewage cause algae blooms, which kill sea grass, which is vital to the ecosystem in places like Biscayne Bay. Some algae blooms are harmful to humans, such as blue-green algae, which is a liver and nerve toxin. Floridians got a vivid demonstration of this with Indian River in 2016, when one of the most diverse estuaries in the United States became a river of flocculent green glop that killed millions of fish and inspired YouTube videos of manatees being saved by Good Samaritans who turned on their hoses to douse them with freshwater. The Indian River glop was caused in part by excessive nutrients from agricultural runoff stored in Lake Okeechobee, but leaky septic systems along the river were also suspected as a prime cause.

The public health risks of drinking water polluted by human waste are well documented. In 2010, some 10,000 people in Haiti died and hundreds of thousands more were sickened by cholera, in an epidemic resulting from the mishandling of septic tanks at a UN Peacekeeper camp and dumping into a river. As late as the 1920s, before the advent of modern septic systems in the United States, typhoid was a fairly common disease.

But put aside drinking it—even swimming or bathing in

contaminated water is risky. According to Erin Lipp, the chance of ingesting bacteria that cause diseases like cholera and typhoid is low, in part because even in soggy septic systems, the bacteria are big enough that they are trapped in the pores of the soil. The bigger risks are enteric viruses, most of which cause fevers, rashes, and diarrhea, but some of which are more serious, such as hepatitis A. And the risk doesn't go away quickly. Enteric viruses can live for weeks in seawater.

The best fix for leaky in-ground septic tanks in a city like Miami is to get people hooked up to a municipal sewer system and then make sure the sewer system is well maintained. But that takes money—and planning. It costs about $15,000 for a homeowner to connect to a municipal sewer line—if one exists in the neighborhood. If it doesn't, the county has to install a new line, which also costs a significant amount of money.

Leaky septic systems are not the only source of contamination urban residents need to be concerned about when their city starts to flood. In Miami, a 200-acre garbage dump known affectionately as Mount Trashmore sits right on the edge of Biscayne Bay. Since it opened in 1980, millions of tons of all the chemical-laden debris have been dumped there—fingernail polish, printer ink, oven cleaner, Freon, motor oil, degreaser, house paint, weed killer, fertilizers, rat poison. The disposal pit is sealed off by layers of clay at the bottom, but this stew of arsenic, chromium, copper, nickel, iron, lead, mercury, zinc, and benzene was never designed to be submerged.

In South Florida, not even the dead are safe from rising seas. Coffins are buoyant, and when water comes, they sometimes rise out of the ground. In other cases, water lifts the lids, allowing the remains to float out. This is not an uncommon event in flooded areas. In 2015, after a few days of extreme rainfall in Baton Rouge, Louisiana, dozens of coffins had to be recovered and reburied.

In Miami, many cemeteries are on low-lying ground that will be quickly inundated. Julia Tuttle, the founder of Miami, is one of the lucky ones. Miami City Cemetery, where she is buried, is 10 feet above sea level. On the other hand, 1950s comedian Jackie Gleason, who is interred at Our Lady of Mercy Cemetery, is entombed only 3.5 feet above sea level. Mount Nebo Miami Memorial Gardens, where gangster Meyer Lansky is laid to rest, is 4.5 feet above sea level. Actor Leslie Nielsen, who is interred at Evergreen Cemetery in Fort Lauderdale, is at 7 feet. Omar Mateen, the shooter who killed forty-nine people in a Florida nightclub in 2016, is buried in a Muslim cemetery that is just four feet above sea level. In the historic Key West Cemetery, about 60,000 people, including former slaves and Cuban freedom fighters, are buried less than eight feet above sea level.

A few days after my adventures with Briceño, I visited Bruce Mowry in his cluttered office on the third floor of Miami Beach City Hall. That morning, as usual, he was going in a thousand directions at once. "I don't think I've put in less than a twelve-hour day since I've been here," he boasted. He was living during the week at a small apartment in mid-beach, while his wife stayed at their permanent home 250 miles north near Daytona Beach (he visited on weekends). With a staff of twelve engineers, it was Mowry's job to keep the city from drowning—and, equally important, to keep people from *thinking* the city was drowning.

Mowry is not a visionary. He is a tactician. He is the guy who designs the plan to get the troops up the beach on D-day. "Basically people don't like me because they feel like I intimidate them, but I tell them, 'Hey, I'm here for one reason.' As I told the city manager when I came here, 'I'm here to build as many projects as I can before I die.' Some people say, 'That's not a real good attitude.' And I say, 'That's the way I operate. I build projects.'"

Here are some of the projects that were on Mowry's to-do list the day we talked: To solve the problem of bacteria in water being discharged into the bay, he was looking at a system that would treat all discharges with ultraviolet light, which would kill bacteria or viruses (this would solve the problem of contamination in the bay but would do nothing to help with leakage from in-ground septic systems, which was the cause of the contamination I saw in Shorecrest, which is not part of Miami Beach). To measure rising groundwater beneath the city, he was drilling forty-two observation wells. To better monitor local sea levels, he was installing two new tide gauges in different parts of the city. He was overseeing the installation of a half dozen new, bigger pumps (Mowry says the city's master plan calls for a total of sixty pumps in low-lying areas; by early 2017, about thirty were installed and operational). He was arguing with historic preservationists about building code requirements in historic neighborhoods. He was talking with other city officials about mandating higher first-floor ceilings, so that as seas rose in the future, the bottom floors of buildings could be raised and there would still be enough headroom on the first floors. He was engaging design firms to figure out clever ways to hide diesel generators for the pumps (people complain that the generators are too ugly and too noisy). His highest-profile project was overseeing a $25-million venture with the state to raise Indian Creek Drive, a key thoroughfare on the bay side of the island, as well as to build a new seawall and install yet another pump station at the same location.

The ultimate goal of all of this, in Mowry's view, was to buy time. "I look at sea-level rise as basically an opportunity to start upgrading our infrastructure, but do it in a common-sense manner. When a road needs to be replaced, go in and replace it—but raise it higher while you're at it. I think by triggering certain things like that, you create a domino effect. In the next thirty to fifty years, this city is going to get higher and higher. Buildings that are cost-effective to raise will eventually be raised. Buildings that aren't will be demolished and something new will be built."

But Mowry understands the need to move fast. "The problem is that we have the economic engine here, and we have to get moving before the economy crashes on this city because of not doing anything. We need to get started *now*, while we have the revenues to do it. You don't wait until water starts pumping out of the ground and say, 'Oh, I guess we need to start looking at groundwater issues.'"

Mowry also understands that engineers are not gods, and that engineering has its limits. What will allow Miami to thrive in the future is not simply the fact that it's not under-water. It will thrive if it is a lively, creative, thriving, safe, and equitable city—and one that also happens not to be underwater.

"From my point of view, there's nothing that I can't over-come," Mowry said. "But I am not the only one here. This is a city. I only play a function as technical advisor, telling civic leaders what I can do and what we can do. But if residents don't agree, if the businesses don't agree, and the visitors coming to this city don't agree, then the economy of the city dies and the engineer's unemployed. In other words, just because I can tell you how to solve the problem, if it doesn't fit within the culture of this city and the future of this city, then it's not the right solution."

What exactly the Miami of the future will look like is a subject of constant discussion among those who understand the risks the city faces. A few weeks earlier, I had attended an evening talk in downtown Miami at the local chapter of the American Institute of Architects, which is headed by architect Reinaldo Borges. At the talk, landscape architects Walter Meyer and Jennifer Bolstad of Local Office Landscape Architecture outlined a plan to reshape a Miami neighborhood that was, in the context of Florida's pave-the-swamp development mentality, both subtle and subversive. Meyer and Bolstad called their approach "forensic ecology." Like the innovative proposals by Susannah Drake and Kate Orff to protect New York City that I mentioned in an earlier chapter, Meyer and Bolstad aimed to work with nature, not against it. As an example, they showed plans for the redevelopment of Arch Creek Basin in north Miami, where they proposed tearing down houses that had been built in a low-lying neighborhood that had been a natural slough, connecting the Everglades to the sea, and building new housing on higher ground. "The idea is to reduce risk by relocating people, but doing it locally, so that families and neighborhoods are not disrupted," Meyer said. The natural slough would be restored, allowing water to collect and drain along the natural contours of the land. This was not a long-term solution to sea-level rise for the neighborhood, but it would buy time and encourage city planners to think about the natural landscape as something more than just a pile of dirt to be bulldozed and paved over.

After the event, I walked out with Borges, and we sat for several hours in his Range Rover Sport in the parking lot, talking. (Borges says his Range Rover is the perfect car for Miami: "It can roll through four feet of water, no problem.") During the course of reporting this book, I spent many hours with Borges and found him to be deeply thoughtful on the issue of sea-level rise. As an architect, he understands the imperative to keep the economic engine of Miami humming. As a father of two daughters, he also understands the need to push city officials to prepare Miami for a watery future. He has spent hours in meetings, arguing for better building codes and pointing out the insanity of building condo towers with underground parking lots. But he also understands the draw of living by the ocean (Borges himself lives on the twenty-fifth floor of a condo tower in downtown Miami). "People love water," he told me. "They love the sense of peace it gives them, the restfulness. If I could build a building that could just hover over the water, people would love it."

Borges views sea-level rise as a creative challenge, and one

Illustration of Reinaldo Borges' platform city in Miami's Biscayne Bay. *(Illustration courtesy of Reinaldo Borges)*

that is only limited by our own imaginations. Or as Harvard philosopher Roberto Mangabeira Unger put it: "At every level the greatest obstacle to transforming the world is that we lack the clarity and imagination to conceive that it could be different." Borges is enamored of Japanese architect Kiyonori Kikutake, who worked out elaborate plans for floating cities in Tokyo Bay in the 1950s. In his spare time, Borges had been sketching out his own ideas, including building water-view condos on both sides of the bridge to Virginia Key, a notion inspired by the Ponte Vecchio, the famous medieval bridge in Florence. Another idea, which occurred to him when he saw a picture of an oil rig, was to construct a series of platforms in Biscayne Bay, elevated seventy-five feet or so above the water on heavy pilings, each holding high-rise towers for thousands of people and accessible by water ferries. "Why can't we do something like that?" Borges said that night as we sat in the dark parking lot. "Yes, we have a lot of problems, and yes, it will require some radical thinking. But I think this is an exciting time. One of the great strengths of Miami is that it's still a new city, still growing, still forming its identity. There is so much energy and money and creativity here. We just need to put it to work in a new way."

In Miami, as in every other city in the world, there is hope that if sea levels rise slowly enough, it will erode the politics of denial and inspire innovation and creative thinking, and the whole crisis will be manageable. People who don't want to live with the risk of higher water can move to Denver, while the people who want to experiment with a platform city and new ways of living with water can remain behind, pioneers in a watery urban renewal.

The problem is, rising seas also raise other kinds of risks—including the risk of sudden catastrophe. In Miami, the biggest concern is a potential nuclear disaster at the Turkey Point nuclear power plant, which sits on the edge of Biscayne Bay just south of Miami, completely exposed to hurricanes and rising seas. "It is impossible to imagine a stupider place to build a nuclear plant than Turkey Point," said Philip Stoddard, the mayor of South Miami and an outspoken critic of the plant.

The Turkey Point nukes began operation in the early 1970s, long before sea-level rise was a well-recognized risk. But precautions were taken to protect the plant from hurricanes; most important, the reactor vessels are elevated twenty feet above sea level, several feet above the maximum storm surge the region has seen. According to Florida Power &

"It is impossible to imagine a stupider place to build a nuclear plant than Turkey Point," says Philip Stoddard, the mayor of South Miami. *(Photo courtesy of Shutterstock)*

Light, the electric utility that operates the plants, there is virtually no chance of a storm surge causing problems with the reactors. As evidence of this, Michael Waldron, a spokesman for the company, points to the fact that Hurricane Andrew, a Category 5 hurricane, passed directly over the plant in 1992, with very little damage. "It goes without saying that safety is our number-one priority," Waldron said in an email.

But Stoddard and other critics of the plant are not reassured. For one thing, although the plant did weather the hurricane, the peak storm surge, which was seventeen feet high, passed ten miles north of the plant. According to the late Peter Harlem, who was a noted research geologist at Florida International University, the plant itself only weathered a surge of about three feet—hardly a testament to the storm-readiness of the plant. How would Turkey Point fare if it was hit with a Hurricane Katrina–size storm surge of twenty-eight feet?

Stoddard also pointed out that although the reactors themselves are elevated, some of the other equipment is not. "I was given a tour of the plant in 2011," he said. "It was impressively lashed down against wind, but even I could see vulnerabilities to water." Stoddard noticed that some of the ancillary equipment was not raised high enough. He was particularly struck by the location of one of the emergency diesel generators, which are crucial for keeping cooling waters circulating in the event of a power failure (it was the failure of four layers of power supply that caused the meltdown of reactors in Fukushima, Japan, after the plant was hit by a tsunami in 2011). Stoddard said the generator was located about fifteen feet above sea level, and it was housed in a container with open louvers. "How easy would it be for water to flow into that? How well does that generator work when it is under water?"

Another problem: Turkey Point uses a system of cooling canals to dissipate heat. Those canals are cut into coastal marsh surrounding the plant, which is only about two feet above sea level. Besides being vulnerable to storm surges, the cooling canals are also polluting the bay. In 2014, regulators discovered that the leaky canals were pushing an underground plume of salt water miles inland, threatening drinking water supplies, and leaking water tainted with tritium, a radioactive element, into the fragile waters of the Biscayne National Park.

But the biggest problem of all is that inundation maps show that with just one foot of sea-level rise, the cooling ponds begin to flood, with two feet of rise, they are inundated, three feet of rise, and Turkey Point is cut off from the mainland and accessible only by boat or aircraft. And the higher the seas go, the deeper it's submerged.

According to Dave Lochbaum, a nuclear engineer and the director of the Nuclear Safety Project for the Union of Concerned Scientists, the situation at Turkey Point underscores the backwardness of how we calculate the risks of nuclear power. The Nuclear Regulatory Commission, which oversees the safety of nukes in America, demands that operators take into account past natural hazards such as storms and earthquakes, "but they are silent about future hazards like sea-level rise and increasing storm surges," Lochbaum said. The task force that examined nuclear safety regulations after the Fukushima tsunami recommended that the NRC begin taking future events into account, but so far, the commission has not acted on the recommendation.

Still, Florida Power & Light insists the plant is perfectly safe. When I asked for details about FPL's plans to armor the plant against sea-level rise, its PR reps were elusive. They told me the plant's current design is suitable to handle sea-level rise but would not tell me how much. (Six inches? Six feet?) They would not disclose plans to protect or redesign the cooling canals. They assured me that "all equipment and components vital to nuclear safety are flood-protected to twenty-two feet above sea level." But when I asked to visit the plant and see for myself, they refused.

I went out there anyway. I was denied access to the inner workings, but I got a very nice view of two forty-year-old reactors perched on the edge of a rising sea with millions of people living within a few miles of the plant. It was as clear a picture of the insanity of modern life as I've ever seen.

Florida Power & Light thinks Turkey Point is such a great place for nukes that it has proposed a $20-billion plan to build two more reactors out there. Given the life expectancy of a nuke plant, it means that the people of South Florida would likely live with the threat of a radioactive cloud over their heads until at least 2085. In late 2016, after a seven-year environmental review, federal regulators approved the plan, stating that the two new reactors would have virtually no impact on the environment. The report said nothing about sea-level rise.

12. THE LONG GOODBYE

A DECADE AGO, an elite group of scientists, economists, and government officials gathered at Snowmass ski resort near Aspen, Colorado, to contemplate the end of the world. The week-long workshop, held in the shadow of fourteen-thousand-foot peaks at the Top of the Village lodge, was organized by the Energy Modeling Forum, a group of academics and industry leaders affiliated with Stanford University. A few months earlier, Stanford professor John Weyant, the director of the group, had asked participants to consider a nightmare scenario: It's a decade or so in the future, and the impacts of climate change are accelerating. The Greenland and Western Antarctic ice sheets are melting at an exponential rate, leading to predictions of a dramatic rise in sea levels by 2070. In this scenario, southern Florida vanishes, New York City becomes an aquarium, London looks like Venice. In Bangladesh alone, 40 million people are displaced by the rising waters.

If you needed to put a "sudden stop" on emissions of CO_2, Weyant asked, how—short of shutting down the global economy—would you do it?

At the Snowmass workshop, it was clear that putting a "sudden stop" to climate-warming emissions would require something more than investing in wind turbines. In one presentation, Jae Edmonds, chief scientist at the Pacific Northwest National Laboratory, suggested that one way to radically cut emissions without shutting down the economy would be to capture and bury CO_2 emissions from power plants, as well as to replace coal and oil with genetically engineered biofuels, which could potentially create negative emissions if the crops used as a fuel source suck up enough carbon dioxide as they grow. But accomplishing this would require a massive expansion of agriculture, sweeping changes to the world's energy infrastructure, bold political leadership, and trillions of dollars.

Then Lowell Wood approached the podium. At sixty-five, Wood was a big, rumpled guy, tall and broad as a missile silo, with a full red beard and pale blue eyes that burned with a thermonuclear glow. In scientific circles, Wood was a dark star, the protégé of Edward Teller, inventor of the hydrogen bomb and architect of the Reagan-era Star Wars missile defense system. As a physicist at Lawrence Livermore National Laboratory for more than four decades, Wood had long been one of the Pentagon's top weaponeers, the agency's go-to guru for threat assessment and weapons development. Wood was infamous for championing fringe science, from X-ray lasers to cold-fusion nuclear reactors, as well as for his long affiliation with the Hoover Institution, a right-wing think tank on the Stanford campus. Everyone at Snowmass knew Wood's reputation. To some, he was a brilliant outside-the-box thinker; to others, he was the embodiment of Big Science gone awry.

Wood hooked up his laptop, threw his first slide onto the screen, and got down to business: What if all the conventional thinking about how to deal with global warming was *wrong*? What if you could do an end run around carbon-trading schemes and international treaties and political gridlock and actually *solve the problem*? And what if the cost to get started was not trillions of dollars, but $100 million a year—less than the cost of a good-size wind farm?

Wood's proposal was not technologically complex. It was based on the idea, well proven by atmospheric scientists, that volcano eruptions alter the climate for months by loading the skies with tiny particles that act as mini reflectors, shading out sunlight and cooling the Earth. Why not apply the same principles to saving the ice sheets? Getting the particles into the stratosphere is not a problem—you could generate them easily enough by burning sulfate, then spray the particles out of a few high-flying jets. The particles would fall out of the sky after a few months, so you'd have to keep spraying more or less constantly. They'd be invisible to the naked eye, Wood argued, and harmless to the environment. Depending on the number of particles you injected, you could not only stabilize Greenland's ice—you might actually *grow* it. Results would be quick: If you started spraying particles into the stratosphere tomorrow, you'd see changes in the ice within a few months. And if it worked over the Arctic, it would be simple enough to expand the program to encompass the rest of the planet. In effect, you could create a global thermostat, one that people could dial up or down to suit their needs (or the needs of polar bears).

Reaction to Wood's proposal was fast and furious. Some scientists in the room, including Richard Tol, a climate modeler with the Economic and Social Research Institute in Dublin, found Wood's ideas worthy of further research. Others, however, were outraged by the unscientific, speculative, downright arrogant proposal of this...this *weaponeer*. The Earth's climate, one scientist argued, is a chaotic system— shooting particles into the stratosphere could have unforeseen consequences, such as enlarging the ozone hole, that we might only discover after the damage was done. What if the particles had an effect on cloud formation, leading to unexpected droughts over Europe? Bill Nordhaus, a Yale economist, worried about political implications: Wasn't this simply a way of enabling more fossil fuel use, like giving methadone to a heroin addict? Besides, if people believed there was a solution to global warming that did not require hard choices, how could we ever make the case that they needed to change their lives and cut emissions?

John Weyant, surprised by the "emotional and religious" debate over Wood's proposal, cut off discussion before it could turn into a shouting match. But Wood was delighted by the ruckus. "Yes, there was some spirited discussion," he boasted to me when I had lunch with him at a Mexican restaurant in Silicon Valley shortly after the event. "But a surprising number of people said to me, '*Why haven't we heard about this before? Why aren't we doing this?*' "

Then Wood flashed a devilish grin. "I think a few of them were ready to cross over to the dark side."

We live in a rapidly accelerating technological age, with new iPhones every year that make the old ones seem as primitive as a brick, where robots perform surgery and computers fly 757s. Scientists are unraveling the mysteries of DNA and plotting the circuitry of the human brain. Techno-optimists like Ray Kurzweil talk openly about immortality. Elon Musk aspires to create a "multiplanet civilization" in the very near future. It seems only natural that a slow-moving force like sea-level rise would have a technological solution too. Why *not* build a thermostat for the planet? We are already engineering the Earth's operating system by dumping billions of tons of greenhouse gases into it every year. We're just doing it badly. Why not get good at it?

When Wood made his presentation in Aspen a decade ago, very few people had heard of geoengineering. Now it is openly debated in the climate and energy world. Mainstream scientists are cautiously acknowledging that geoengineering could indeed be an important tool in the what-the-hell-do-we-do-about-climate-change toolbox. As the IPCC's most recent report concluded, "Models consistently suggest that [solar geoengineering] would generally reduce climate differences compared to a world with elevated greenhouse gas concentrations and no [solar geoengineering]." Other scientists, including the United Kingdom's Royal Society and the United States' National Academy of Sciences, support further research. Environmental groups like the Environmental Defense Fund and the Natural Resources Defense Council do too. Near the end of Obama's second term, the White House went so far as to recommend a federally funded research project to better understand the risks and benefits of geoengineering (alas, nothing came of it). But just as awareness of the need for further research has grown, so has awareness of the potential downside. At the 2017 World

Economic Forum in Davos, Switzerland, geoengineering was cited as one of the top risks that the world faces.

And what Wood said ten years ago at that meeting remains true today—when you think about big technological fixes for sea-level rise, spraying particles in the atmosphere to reflect away sunlight is the only planetary-scale fix we know of that could plausibly stop or slow sea-level rise. Other ideas, such as pumping billions of tons of ocean water onto Antarctica, where it would freeze and lower sea levels, or re-creating an ice-age landscape (complete with genetically engineered beasts that would be a cross between elephants and woolly mammoths) in Siberia to help reflect sunlight and keep the tundra frozen, may be provocative thought experiments, but few scientists take them seriously.

Wood was a bit optimistic about the costs of a full-scale geoengineering program—David Keith, a Harvard professor who has thought deeply about how to design and implement a geoengineering program, estimates that instead of costing $100 million a year, it would cost more like $2 billion a year. If that sounds like a lot, consider that global subsidies for the fossil fuel industry are about a thousand times that ($1 trillion a year).

Wood, being the technology-happy guy he is, also downplayed the risks of geoengineering during his talk at Aspen. Although reflecting away some sunlight before it hits the Earth could slow the melting of the ice sheets on the surface, it would be a very long time before this had any impact on the warming of the oceans, which is the most immediate threat to the big glaciers in West Antarctica. Nor would it do anything to reduce ocean acidification, which is caused by high levels of CO_2 in the atmosphere and is already damaging coral reefs and threatening the ocean food chain. There are also risks to the ozone layer, not to mention the fact that people would be breathing in these particles as they slowly fall out of the sky (today about 6.5 million people die prematurely each year from air pollution; Keith estimated that a full-scale geoengineering program might lead to thousands of additional deaths each year, but those deaths would likely be offset by hundreds of thousands of lives saved by reduced heat exposure). Finally, Wood neglected to point out that once we started spraying particles into the stratosphere, we would have to keep it up for decades or risk a sudden warming—creating, in effect, a climate version of the Sword of Damocles hanging over our heads.

Geoengineering also brings up complex questions about governance. Whose hand controls the thermostat? Vladimir Putin, after all, might be happy with higher temperatures than Tony de Brum in the Marshall Islands. You don't have to be a science fiction writer to see how geoengineering—or even the threat of geoengineering—could lead to conflict and even climate wars.

Of course, no one would be talking about geoengineering as a solution to sea-level rise (or anything else) if the world had gotten serious about cutting greenhouse gas emissions thirty years ago when scientists first began to raise the alarm. And just to be clear: despite what some conspiracy theorists may believe, the Illuminati are not spraying particles into the atmosphere right now. Nor are any large-scale field research programs under way. Most of what we know about geoengineering comes from computer models, as well as a few modest laboratory experiments. To better understand how particles react in the stratosphere, Keith and a colleague at Harvard have proposed spraying a few kilograms of particles into the stratosphere from a hot-air balloon above Arizona. In a glimpse at what will surely be future fights over the ethics of geoengineering, critics have already cast Keith's modest and valuable experiment as the first move toward the creation of Planet Frankenstein.

The difficult truth about geoengineering is a seductively simple technological fix for a wickedly complex problem. It doesn't require us to change our lives, or to pay more money for energy, or to swap out our SUV for a skateboard. It just requires people to sign off on the idea of a few airplanes spraying particles high up in the stratosphere every week and agreeing to trust someone else to manage the climate for us.

And that is exactly why it is dangerous. Instead of putting faith in individual action to address the problem of climate change, geoengineering puts faith in the magic of technology. As Yuval Noah Harari, an Israeli historian and the author of *Homo Deus,* put it in an email: "Governments, corporations, and citizens allow themselves to act in a very irresponsible way because they assume that when push comes to shove, the scientists will invent something that will save the day."

I thought about this on a warm spring night in 2015, while I watched Philip Levine, who was then mayor of Miami Beach, welcome people to the celebrity-filled 100th anniversary of the city. The event was held, appropriately enough, on a stage set up right on Miami Beach. Levine is a smart guy, and he didn't hesitate to address skeptics who believed that Miami Beach's 200th anniversary would be celebrated in scuba suits. "I believe in human innovation," Levine told the crowd. "If, thirty or forty years ago, I'd told you that you were going to be able to send messages to friends around the world with a phone you carried around in your pocket, you would think I was out of my mind." Thirty or forty years from now, he said, "We're going to have innovative solutions to fight back against sea-level rise that we cannot even imagine today."

Translation: Party on, folks. The future will take care of itself.

But the future will not take care of itself. It will be shaped by decisions we made yesterday and will make tomorrow. For people who live in coastal cities, dealing with sea-level rise will require a lot of difficult choices, including in which neighborhoods to invest in new infrastructure, where to build seawalls, which historic structures to save and which to let go ("You can only save so many lighthouses," Lisa Craig, the chief of historic preservation in Annapolis, Maryland, told me). Smart cities will develop master plans, articulate long-term strategic visions, revise zoning ordinances, pass tax incentives to shift development to higher ground. But that's just a start.

Of all the hard decisions people who live on vulnerable coasts will have to face, the most difficult one is the idea of retreat. Retreat, after all, is what you do if you're standing on the beach and the tide comes up too fast. You get out of the way. It's what the Calusa did on the coast of Florida a thousand years ago, and what the hunter-gatherers who lived in in the now-flooded lands beneath the North Sea did ten thousand years ago. But we modern humans have poured a lot of concrete and asphalt and erected a lot of steel on the beach, and that makes it far more difficult for us to just fold up our tents and move to higher ground.

In many ways, retreat is the opposite of geoengineering: instead of relying on scientists to take care of the problem for you, it requires individual action, thoughtful planning, and a willingness to change your life. Most of all, it means giving up the war with water and admitting that nature has won. That is not a feeling many people embrace. In strictly practical terms, retreat also requires city and state officials to willingly shrink their tax base and politicians to willingly give up power. Who wants that?

Consider the case of Toms River, New Jersey, a town of 92,000 people along the Jersey shore about seventy-five miles south of Manhattan. Until the 1950s, Toms River was a quiet, rural place, best known for the egg farmers who shipped their goods off to New York City. Then the chemical industry arrived and built dye factories. Population boomed for a few decades; then the factories shut down, made obsolete by innovation and global competition, leaving a toxic legacy of cancer clusters and Superfund sites and a lot of cheaply built houses on the beach and along Barnegat Bay, which separates the barrier-island beaches from the mainland.

In some ways, Toms River resembles a working-class version of Miami. The heart of the town is on the shore, but the soul of the town is across the bay, at the beach, which, like Miami Beach, is a wispy island of sand facing the Atlantic. Like Miami, Toms River depends heavily on beach tourism, and like Miami, the town is extremely vulnerable not just from the ocean side, but also from the bay side, where many homes were built right on the water. Already Toms River has one of the highest rates of nuisance flooding on the Jersey shore. With sea-level rise, it will only get worse. According to a report by the Regional Plan Association, an influential group of industry leaders and university researchers in the New York area, one foot of sea-level rise will inundate 3,000 residents around Barnegat Bay. With three feet, 23,000 residents will be underwater. "At six feet of sea-level rise," the report concludes, "the story of the Jersey shore is the loss of arcades, boardwalks, amusement parks and sands that fuel New Jersey's tourism economy."

Toms River was hit hard by Hurricane Sandy. A nine-foot storm surge had inundated the town, damaging or destroying 10,000 homes. On Ortley Beach, a neighborhood out on the barrier island, all but 60 of the 2,600 homes were damaged or destroyed. Houses on the backside of the island, as well as along the bay, were protected from the worst of the surge but were still flooded by rising waters in the bay. Miraculously, nobody in Toms River was killed. A drowned roller coaster in Seaside, a nearby town, became one of the iconic images of the disaster.

In the aftermath, there was much talk about how to reduce risks from future flooding events. A team of scientists and researchers at Rutgers University spent a year talking with local officials and members of the community to come up with a plan, as one document put it, "to help to shift the barrier island communities away from an over-reliance on the beach toward a more nuanced, diverse, sustainable relationship to the shore." The Rutgers team wanted to create an inland "pier" or passageway to connect the coast with the nearby Pine Barrens, a heavily forested area with a unique coastal ecosystem (orchids and carnivorous plants), allowing for easy movement of people and wildlife. They imagined connecting the beach with inland areas by means of new, more sea-level-rise-friendly transportation systems, including aerial trams and water taxis. But they also imagined that as the seas rose, beach tourism would give way to a broader and more sustainable kind of ecotourism, including hiking and biking and bird-watching in the Pine Barrens. The plan included five thousand new housing units on higher ground to ease the transition away from the coast.

All in all, it was a bold vision, and one that would take a lot of time, money, and political leadership to achieve. But it would have begun transforming the city into a place that might thrive in a world of rising seas and increased storms.

Instead, Toms River was rebuilt exactly the way it was. Well, not *exactly*: most of the rebuilt homes were elevated several feet, and critical infrastructure like electric panels were moved out of the way. In addition, the Army Corps of Engineers agreed to spend $150 million to construct a stronger dune for several miles on the Atlantic side of the barrier island, including the section in front of Toms River. The Army Corps has been building "reinforced" dunes like this since the 1960s, and they offer some protection from storm surges for a while, but need to be constantly rebuilt. It might be good for the Army Corps too, because it gives them purpose and allows the agency to ask for a bigger chunk of the federal budget every year, but in the long run, these dunes are just sandcastles on the beach. As far as sea-level rise goes, a higher dune does nothing to protect residents from rising waters in Barnegat Bay, which is where the real risk is. Yes, a few wealthy residents had turned their homes into elevated fortresses, but all in all, five years after Sandy had rolled in, Toms River was not in much better shape to face the future than it had been before the storm.

Why? The simplest explanation is that people in Toms River like where they live and don't want it to change. As one longtime resident told me, "This is a unique place, the best of small-town America. I love it just the way it is." And sea-level rise? Most of the people I talked to were far more worried about radical Islamic terrorists than climate change. When I asked Mayor Tom Kelaher, an eighty-three-year-old three-term Republican, about his views on climate change, he said, "I think the climate is changing, but whether it is

Toms River, New Jersey. Damage from Hurricane Sandy, top, and after rebuilding, bottom. *(Photo courtesy of Christopher J. Raia/Toms River Police Department)*

caused by human beings or not, I can't tell you." When I asked him if the majority of residents of Toms River felt the same way, he said, "I don't know. It's not something we talk about a lot around here." I suspected that Mayor Kelaher might have stronger views about climate change than he was willing to let on (a few years ago he traveled to Norway as part of a group to learn about climate science) but that he understood that it was dangerous to talk openly about it. After all, Toms River voted two to one in favor of Trump in the 2016 election. Kelaher knows as well as anyone that if you thought climate change was a serious problem, you probably didn't vote for Trump.

But it wasn't just ideology. It was also money. About $2 billion in taxable property in Toms River was destroyed in the storm, causing a shortfall of $18 million in the city's annual budget. For a city like Toms River, which thrives on low taxes, this was a crisis. If the city increased taxes, Kelaher and others believed residents would flee. If it didn't raise taxes, however, it would have to cut services.

In the end, the city was forced to raise taxes a small amount. But mostly they tried to increase the tax base by encouraging people to build back bigger homes. In some cases, instead of strengthening building codes and changing zoning regulations to motivate people to move out of risky areas, the city loosened them. New Jersey officials also worked to make sure that FEMA didn't make any big changes in the floodplain designations for the town. When I asked the mayor if flood insurance rates had gone up after Sandy, he said, "Not really."

This is how disaster relief works in America. There are lots of incentives to rebuild but few incentives to rebuild differently, much less to rethink the long-term future of cities and towns along the coast. And it's not the people who live in town like Toms River who pay the costs anyway, so they have no incentive to change their thinking. By the end of 2016, the State of New Jersey had spent $4.6 billion on Sandy recovery efforts, 95 percent of which came from the federal government. In effect, people in Kansas and Washington and Iowa—people who will probably never see a Jersey beach—paid for the reconstruction. In Toms River, where four thousand homes were substantially damaged, US taxpayers subsidized recovery efforts with about

$300 million in federal recovery funds. The town was granted another $30 million in state funding over five years simply to help close the budget gap caused by lost property tax revenues.

Given the risks that towns like Toms River face in the future, you have to wonder: How long is this sustainable? As seas rise and flooding gets more and more frequent, damaging, and costly, more and more cities and towns will be begging for more and more help, basically arguing, as New Jersey officials did, *If you don't bail us out, this town dies.* "The question is, when do taxpayers who are not benefiting from this wake up and say, 'We're not paying for this'?" said Peter Byrne, director of the Environmental Law and Policy Institute at Georgetown University. "There has to be a limit to how long the public will pay for protecting beachfront property when we all know it is going underwater anyway."

During one of my visits to Toms River, I took a walk on the boardwalk with Mayor Kelaher. A few days earlier, a nor'easter had blown in, wiping out a temporary dune the town had built while they were waiting for the Army Corps of Engineers to construct something more substantial. While Kelaher and I talked, dump trucks lined up near us, bringing sand to the shore to protect the beachfront houses in case they were blasted by another storm. Kelaher told me the two thousand truckloads of sand would cost the city nearly a million dollars when all was said and done. "But what are we gonna do? We need some protection from the waves," he said, standing on the boardwalk, his green knit tie blowing in the cold winter wind.

He pointed to the rows of neatly kept beach houses. "People love it here," he said with obvious pride. "They have been bringing their families here for years, every summer. They are very attached to it."

"Do you think sea-level rise is a risk for this town?" I asked.

"Oh, I think it is. But not in my lifetime."

I pointed out, gently, that he was eighty-three, so that wasn't saying much. He laughed.

"What would happen if you told people here that, even with the dune, the risks of flooding are going to get higher and higher in the coming decades, and that if they aren't ready for that, maybe they should sell their houses and move to higher ground?"

Kelaher looked as me as if I were crazy. "If I went door to door in this neighborhood, telling people that, I wouldn't get out of here alive. If you told people they couldn't live here anymore, it would be an economic and emotional catastrophe."

In a world of quickly rising seas, the rationale for encouraging people to move out of harm's way is straightforward: it saves money and it saves lives. For elected officials, the rationale for *not* encouraging people to move out of harm's way is also straightforward: if you ask voters to do something difficult, something time-consuming, or, worst of all, something that costs them money, you get voted out of office. Or sued.

When it comes to new development, many cities have developed incentives to encourage people to build on higher ground. Zoning ordinances and restrictions on how close you can build to the water are the simplest. Some cities, including Miami, are considering levying impact fees on developers, especially if they want to build in low-lying areas, then using the money to fund clean energy and climate adaptation projects.

But the most difficult problem is not new development; it's the stuff that's already built. Tax incentives and other regulatory tools can encourage people to move, but that is slow and uncertain. Raising flood insurance rates to better reflect the true costs of living in risky places can help. But the simplest way to get people to move out of low-lying areas is simply to buy them out. States can condemn properties with the power of eminent domain, but that is a hostile move that reeks of Big Brother and leads to expensive courtroom battles. Voluntary buyouts avoid all that. In most cases, the state or federal government simply offers to buy out residents at more or less full market value. For people who live in risky areas, where their homes might not be marketable, this is often an attractive option. The State of New York spent $240 million to buy out 610 properties, mostly in a few neighborhoods on Staten Island that had been hard hit by Hurricane Sandy in 2012. In 2016, the US government spent $48 million to resettle twenty-three families who lived on Isle de Jean Charles in Louisiana, which had lost 98 percent of its land to flooding. As waters rise and the risk of flooding increases, pressure will surely build on politicians and civic leaders to find ways of moving people out of harm's way.

As a long-term strategy, however, there are several problems with buyouts. First, they work best when entire neighborhoods volunteer to go. Holdouts force towns to keep providing municipal services (garbage pickup, water, sewage, road maintenance, snowplows, streetlights, police, firefighters) to a shrinking tax base. And they prevent the neighborhood from returning fully to nature, which is often one of the goals.

Another problem is money. Buying out twenty-three families in Louisiana is one thing; buying out an entire town or city is something else entirely. To buy out the 4,000 homes in Toms River that were damaged by Sandy would cost roughly

$1 billion. According to the 2107 Coastal Master Plan for Louisiana, 24,000 homes in the state are at risk for flooding in the next fifty years (and that's assuming the state spends billions to improve dikes and levees and other flood control barriers). Back-of-the-envelope price tag to buy out those homes: $6 billion.

The third problem is, who chooses which homes get bought out and which homes are left to flood? In 2015, Alaskan officials lobbied the US Department of Housing and Urban Development for $62 million in federal funds to help 350 residents in Newtok, a village about five hundred miles west of Anchorage that is rapidly being eaten away by the sea, relocate to higher ground nine miles inland. When President Obama visited the state in 2015, he spoke bluntly about the dangers of sea-level rise. But when it came to doling out money to pay for relocation, HUD officials funded Isle de Jean Charles but not Newtok. Why? HUD officials explained it to *Bloomberg View* writer Christopher Flavelle with vague references to Isle de Jean Charles's "leverage" of local and state funds. Flavelle cited HUD's decision-making process as a good example of "the insane bureaucracy of funding climate adaptation."

As I saw in Toms River, the biggest issue any relocation strategy will have to overcome is simply that people love their homes and don't want to leave. That's less of a political problem in places like China, where more than a million people were forcibly relocated to make room for the Three Gorges Dam. But even in the Netherlands, where there is a strong consensus among the citizens that government has the authority to take whatever actions are necessary to reduce the risks of flooding, relocation is difficult. In Nijmegen, the oldest city in the country, the Dutch government spent nearly $500 million to reroute part of the Waal River to give it more room to spread out and reduce the risk that it would jump its banks and flood parts of the city and surrounding farmland. About fifty residents had to be relocated. "It was not easy," Mathieu Schouten, an advisor to the City of Nijmegen, told me when I visited. "We had to buy everyone out, and offer them new land in a higher place. And even then, a few people didn't want to go. The negotiations were difficult. In some cases, it took eleven years of talking."

Even in Staten Island, where 145 families in the Oakwood Beach neighborhood banded together to convince New York State officials that they should buy them out after Sandy destroyed their homes, I found people who wouldn't budge. When I visited the neighborhood in 2016, many of the homes had been razed and the lots were scrubby meadows. There were still a few abandoned houses that had not yet been torn down, but you could feel nature making a comeback. The streets were lined with *Phragmites australis,* an invasive reed that creates a dense four-foot-high wall of green. Ducks wandered through the streets. Seeing a neighborhood being reclaimed by nature provokes a spooky feeling, like watching the reel of civilization play backward. I found two or three houses that were still occupied. I stopped at one of them—a simple white bungalow with a porch crowded with plastic gnomes—and talked with Lois Kelley, a kind-looking woman in her late fifties. She told me that during Sandy, she had been home alone. As the water came into her living room, she climbed onto the couch. The water kept rising until it was five feet deep. She watched her refrigerator float away, her chairs bobbing in the dark water. "I spent the whole night in the dark, floating on the couch with my cats," she told me. In the wind, she heard the voices of her neighbors, cries for help (she later learned that across the street, a man and his son had drowned in their basement). The next day, the water receded. She was cold and wet, but she was not hurt. She fed her cats, summoned contractors to repair her soggy house (flood insurance covered it all), and went on with her life. A year or so later, when city officials knocked on her door and offered to buy her house at full market value, she refused. "This is my home," she told me. "Why would I leave?"

From a purely economic point of view, it may make sense to spend $3 billion or $4 billion to build a wall around Lower Manhattan. It's more difficult to justify $3 billion or $4 billion to save Hunts Point in the Bronx. Other factors go into deciding what to protect and what not to protect than a blunt cost-benefit analysis (race, historical value, political influence), but the larger point is obviously true: not everyone is going to be saved. Wealthy people will take care of themselves, either by moving their homes or elevating them or building seawalls or simply writing off the house as it crumbles into the sea, but for the vast majority of people who live on coastlines, it's going to be a tough day when they wake up and realize that their state or federal government doesn't have the money or the political will to rescue them.

In the United States, if someone owns property on the beach, it is their right to live there until the land vanishes beneath the waves—at which point, according to common law, the land (now underwater) becomes part of the public trust. But beyond that, the legal questions get more complex. Do you have the right to build a seawall on your property even if it causes flooding for your neighbor? If half your property is underwater, do you pay only half the taxes? If you

live on a high spot in the middle of a swamped neighborhood, do you still have the right to expect essential services,
including fire and police, from the city?

The law is murky on a lot of this, but you can see one version of how it might play out in Summer Haven, Florida, an
old-money enclave just south of St. Augustine in St. Johns
County. Until the 1920s, there were no roads there—people
arrived by boat from St. Augustine or drove along the beach.
The state finally built a road along the coast, designated as
State Route A1A. It was originally made of brick, but after it
got washed out several times, it was rebuilt with gravel and
asphalt. Summer Haven remained a small, exclusive community of less than 100, but as the years passed, residents got used
to the road—it provided the easiest access to their homes.

In the 1970s, state officials grew tired of rebuilding the
road after every storm and moved A1A farther inland; it
turned over about one and a half miles of the old road—
now known to locals as Old A1A—to the county to maintain.
Old A1A became essentially a long driveway for Summer
Haven residents. In 2004, the road washed out again, and
county commissions faced a choice: repair the road for $1
million and see it get washed out again, or leave it alone and
doom the residents to living without road access. The county
had already spent more than twenty-five times the average
annual maintenance costs to keep the road open and in
good shape.

And sea-level rise was just making it worse. The county
decided to build a six-foot-wide berm—using $950,000 in
federal funds—to protect the road from the ocean, but the
berm washed away quickly.

In 2008, after more storms and washouts, the residents of

Homes along Old Highway A1A in Summer Haven, Florida. *(Photo courtesy of St. Johns Public Works)*

Summer Haven had had enough: a group of sixty-five property owners sued the county for failing to maintain the roadway in usable condition. In effect, they argued that it didn't matter if sea-level rise and increasing storm surges were making the road vulnerable—the property owners paid taxes, the county owned the road, and it was their job to maintain it. The property owners argued that the county's failure to do so violated their right to keep existing public access to their properties and therefore constituted a "taking" under the Fifth Amendment of the US Constitution, which states that "private property [shall not] be taken for public use, without just compensation."

This was a highly unusual argument. There have been plenty of cases where property owners have successfully argued that faulty engineering on the part of local, city, or federal officials destroyed their property values. To cite just one example: after Hurricane Katrina, a group of residents argued that a levee designed by the US Army Corps of Engineers had been poorly designed and contributed to the flooding that destroyed their homes. The case went all the way to the US Supreme Court, and the property owners eventually won.

But this case was different. This was not a case of government taking action that damaged their property; the wealthy Summer Haven residents argued that they were harmed because the government hadn't taken *more* action than just spending about twenty-five times as much per mile to maintain their piece of road than the county spent on other roads.

In response, the county argued that as a low-lying coastal road, Old A1A is subjected to continuous damage from natural forces, such as storms and erosion. According to county officials, the only feasible way to protect the road from the "ravages of the ocean" was to spend more than $13 million to elevate the road and then protect it with an embankment of sand that stretched all the way down to the high-tide line. County officials argued it would have to spend an additional $5 to $8 million every few years to maintain that protection. That was more than the entire county budget for repair and maintenance of eight hundred miles of roads in the county. If they had to continue to maintain the road, it could bankrupt the county.

The local court ruled in favor of the county, but the property owners appealed to the district court, which overturned the trial court and found in favor of the property owners, essentially ruling that the county had a duty to "reasonably maintain" and repair Old A1A in such a way as to result in "meaningful access." More significantly, however, the court held that "governmental inaction—in the face of an affirmative

duty to act—can support a claim for inverse condemnation."
This case, for the first time in Florida, established a precedent
that government *inaction* may be grounds for a plaintiff to
bring a constitutional takings claim.

In 2014, property owners and the county worked out a
settlement, with the county agreeing to make "good faith
efforts to preserve, protect, and maintain" the road. But the
court's decision supporting the takings claim still stands.
And for anyone thinking about the economic implications of
sea-level rise to cities and towns, it sets a disturbing precedent. As Thomas Ruppert, a lawyer and coastal planning
expert with Florida Sea Grant, a university-based partner-
ship with state and local governments, wrote in an analysis of
the decision: "If a court were to determine that local governments must maintain a 'level of service' for drainage rather
than maintain the existing infrastructure itself, this could
force local governments into the difficult choice between
spending what may amount to unrealistic sums of money to
keep everyone's property dry or risk facing legal liability."

To state it more bluntly, this means that property owners
in Florida can invoke the Fifth Amendment and sue their
city and town every time a road or bridge washes out due to
flooding—not because the road was poorly designed, but
because when it was built, engineers failed to anticipate sea-
level rise. "It allows residents to tell their local officials, 'We
think you should rebuild our bridge again, and this time
make it twenty feet higher. If you don't do it, we will sue the
government and get paid,'" said Peter Byrne of Georgetown
University. "Thanks to this decision, that mentality is metas-
tasizing in other states." It also means that wealthy, politically
connected residents will be able to dictate how and where
cities and towns spend their money, which will of course

mean less spending for poor neighborhoods that don't have the means to threaten lawsuits.

"As seas rise, cities and towns could go broke trying to maintain their roads and bridges and other infrastructure," said Ruppert. "It's just that simple. And legal decisions like this one could make that day come sooner because they take away the ability of local governments to exercise discretion in how best to serve their citizens with the limited money they have." As cities and towns pull away, Ruppert sees a future coastline emerging that is a mix of watery slums, where people live beyond the reach of government, and enclaves of the superwealthy living in fortresses on the sea, moving in and out with boats and helicopters. "I admit it's apocalyptic," he said. "But that may be where we are headed unless we learn to make difficult, expensive, and painful decisions about adaptation before the crisis hits."

One day while I was exploring the Jersey shore, I drove by some summer cottages in the town of Lavallette, a barrier-island town just north of Toms River. It was winter, and the cottages were all closed up, but I turned off the main road and parked my car and walked around in the neighborhood for a few minutes. Most of the cottages had been built in the 1940s and 1950s, perhaps purchased out of a Sears catalogue and assembled on-site. Some had been fixed up, with second floors added or bigger porches, while others looked like nothing had changed in fifty years. There was no sign of money here. These cottages were not owned by hedge fund managers. They were houses of plumbers and schoolteachers and highway workers who had worked hard and saved hard and got a little place at the beach for their families. Maybe they hung out for two or three weeks a year there, a brief respite from fifty- and sixty-hour workweeks that drove them the rest of the year. And of course many of these houses were built right on the sand.

When I got back to the car, I drove north toward Asbury Park. I've never lived on the Jersey shore, but I've spent enough time there (and listened to enough Bruce Springsteen) to feel an emotional connection to the place. I remembered the night in Atlantic City many years ago when my father phoned me to tell me he was dying, and another, happier night even longer ago, playing Whack-A-Mole with a girlfriend in Wildwood. I hadn't thought of either of these moments in a long time, but our memories adhere to places in surprising ways. As seas rise, we will not only lose homes and property; many of us will lose parts of our past. Others will lose more.

As I drove, I tuned the car radio to a New Jersey NPR station, hoping to catch the news. What I heard instead was a desperate man's voice: "...Well, yeah, after Katrina we lost everything and the state came up with this 'Road Home' program, where they'll buy your property out at a lesser amount than what it's valued at, but if you stay and rebuild they'll give you the full amount. With the understanding that we would get protection and get help. So we built this home here with the understanding that we was gonna get some protection. We got nothing."

I later learned that the voice on my radio belonged to Anthony Caronia, who lived on the shores of Lake Pontchartrain, just north of New Orleans. I had been there a number of times, and as he talked, I could picture the flat bayou landscape, punctuated by strip malls and car dealers. Caronia was being interviewed for a story about Louisiana's $50-billion plan to save the state from disappearing beneath the sea. After

Katrina, the state had encouraged him to stay and rebuild his home. Now he was regretting that he'd agreed.

"...So yeah, the state basically told us to stay, and it would help us out and do all this protection, and nothing—nothing— nothing was ever done. So please, could y'all follow up, the newspeople? And make sure that this doesn't go to the way- side? 'Cause I have five children and a granddaughter and a wife, and I don't expect to live forever but this is their equity, this is gonna be left, this is what I'm—my plan is to leave it for them."

In Caronia's voice, you could hear anger and disappoint- ment and fear: this was a man who had finally realized how much he had to lose.

"But man, the flooding's getting worse and worse and worse and worse. I'm ready, man. I—I—I should have left after Katrina but my wife said 'No, let's—let's just build a house.' It's just a matter—I'm tired. I'm ready to go. The school buses can't turn around on certain days, you know I had to track my kids through water. I mean I had—the National Guard can't even get back to us, the water gets so high. The fire department can't get back there. You're trapped in your house 'cause there's only one way in and one way out. My tractor, my boats, my four-wheelers, my zero-turn lawn mower. Every spring I gotta start sweating, what am I gonna do with my tools?

"I mean, I'm being honest with you, I'm giving up! I'm fifty-one years old—I don't care anymore. And I shouldn't feel that way! This is not right. This is not fair. Something needs to be done today. Today. Please understand me— this is a cry out for help. From anyone and everyone in Amer- ica listening, Mr. Anthony Caronia is begging the State of Louisiana and the United States government to come in and buy me out and please move my family outta harm's way. Please understand my cry. I'm ready to go. I'm begging for help. I'm not asking for help, I'm begging for help."

Shortly after that, the interview ended. I pulled off into a parking lot near the beach and made a note to myself to track down Caronia and learn more about what had happened to him. I guessed that the anger and fear I heard in his voice were just the beginning of a much larger anger and fear that will rise as the water comes and people lose what they love. It reminded me of a conversation I'd had with Dutch landscape architect Adriaan Geuze. I had seen him give a lecture about sea-level rise at a conference in Rotterdam one day, and he'd invited me back to his home for dinner afterward. We sat in his ground-floor kitchen—which was actually below sea level, he pointed out—and talked about the anger and resentment and hardship that are going to surface if our inevitable retreat from the shoreline is not carefully managed. Geuze argued that it will require a radical rethinking of the role of government. If the most basic job of government is to keep people safe, what happens when people realize they are not? What is the government's role in keeping people out of harm's way? Geuze compared sea-level rise to other transformative catastrophes, such as the Dust Bowl of the 1930s, a partly man-made natural disaster that profoundly changed the geography of America and also expanded the role government plays in ensuring the long-term welfare of even the most vulnerable people. "We're going to need a new New Deal," Geuze argued. "It is going to require a rethinking of the social contract between governments and citizens."

Maybe it will. And maybe it will begin with people like Anthony Caronia. After the radio news segment ended, I got out of the car and went for a walk on the beach. It was twilight, and getting cold. I was alone except for a few sea gulls circling overhead. I listened to the crash of waves washing against the shore as they have for millions of years. It sounded like Mother Nature unfurling her sheets and making her bed.

EPILOGUE: CONDO DIVING

In the late 1970s, UC Berkeley geologist Walter Alvarez stopped in Denmark on his way home from a research trip to Italy. Alvarez was hunting for evidence to support his wild and still unproven idea that a planet-wide disaster, such as a giant meteor strike, had killed off the dinosaurs 66 million years ago. The idea that the dinosaurs (and basically all other life on Earth that was larger than a raccoon) could have been killed off by a cataclysmic event went against evolutionary dogma, which held that extinction was a long, slow process, just like evolution itself.

Alvarez, prompted by his famous father, Nobel Prize–winning physicist Luis Alvarez, had other ideas. In Denmark, Walter Alvarez and a colleague drove out from Copenhagen to Stevns Klint (*klint* is Danish for "cliff"), a well-known geological site on the Baltic Sea. The rough white cliffs there are one of the few places in the world where you can see the dinosaurs' last days written in well-defined layers of lime-stone. About halfway up the cliffs a thin, dark line of clay is sandwiched between the limestone layers. "It was clear right away that something unpleasant had happened to the Dan-ish sea bottom where the clay was deposited," Alvarez later wrote. The limestone below the clay line was full of fossils, representing the teeming life of the sea. But the clay bed itself was black, smelled sulfurous, and contained no fossils but fish bones. "During the time interval represented in the clay, the sea bottom had turned into a lifeless, stagnant, oxygen-starved graveyard, where dead fish slowly rotted," Alvarez observed. Alvarez took some samples of the fish clay and discovered that it contained iridium, a rare metal that is scarce on Earth but common in meteors. It was a key piece of evidence in one of the most dramatic scientific discoveries of our time. Today, even third graders know about the meteor that killed off the dinosaurs.

I was lucky enough to spend several weeks in Denmark while I was working on this book, mostly in Copenhagen, a city that prides itself on being one of the greenest in the world. There are bikes everywhere, and about 40 percent of the nation's electricity comes from renewables (the Danish Energy Agency says the grid will be 100 percent renewables by 2035). I met with Lykke Leonardsen, a climate advisor for the city, who told me about some of the measures the city is taking to help drain its streets after increasingly intense rain-falls, such as building "water squares" similar to what I'd seen in the Netherlands. To protect the city from sea-level rise, city planners are already considering a barrier across each end of the main canal, closing it off from the Øresund, the strait that separates Denmark from Sweden and connects the city to the North Sea and the Baltic. In fact, Denmark is more likely to have problems caused by the flood of climate refugees from Bangladesh or Nigeria than by a flood of water from the Baltic Sea. The question of how to absorb refugees from the Syrian war is already tilting Danish politics to the right, and it's hard to see how waves of climate refugees won't accelerate that.

From Copenhagen, it's about an hour and a half's drive to Stevns Klint. Of course, given the subject of this book, I had to go see it. The highway took me out through the suburbs, then into a landscape dotted with well-kept farmhouses, some with thatched roofs. I parked my rental car at the top of a bluff, near an eleventh-century church overlooking the Baltic. A few years ago, as the cliff eroded, the back of the church tumbled into the sea. The rest of the church now hangs there precipitously, awaiting the same fate.

A long flight of metal stairs stretched down to the beach. It was like a staircase into Jurassic Park. One hundred million years ago, when the base of these cliffs was the bottom of the ocean, the world was a very different place: Europe was much closer to America, the seas were three hundred feet higher than they are today, and, of course, dinosaurs roamed the Earth.

I wandered along the curved shore, close to the chalky cliffs, looking for the fish clay. It wasn't easy to spot. But I eventually found it — a thin, broken black layer, like a line of Magic Marker, about forty feet up. It was hard to believe that this was the last and best evidence of one of the most traumatic events we know of on Earth. The meteor that crashed into what is now the Yucatán peninsula in Mexico was the size of Manhattan, and when it hit the Earth, it released the energy of 100 million megatons of TNT. The fire incinerated everything within a thousand miles. Droplets of hot sulfuric acid and dust blocked out the sun. Then it became cold and dark for thousands of years.

It was sunny and mild when I walked on the beach, with a crisp breeze blowing off the Baltic. Flint pebbles rolled under my feet. If you were a dinosaur, I thought, this line of fish clay represented some very bad news. Dinosaurs dominated the Earth for millions of years, but they did not have the traits needed to adapt to a fast-changing world. Long before, changes in the climate, driven by volcanic eruptions, had probably stressed many species and made them vulnerable. The meteor finished them off. But it's also true that for you and me and everyone we know, that meteor strike was a very fortunate event. It was an evolutionary reset, one that allowed mammals to thrive. It is very unlikely that we humans would have appeared and built a city like Miami, much less have submerged it by burning fuels made from decayed plants and animals, if the meteor had not killed off the planet's reigning predators.

Of course, the dinosaurs did not have computer models to help them understand the likelihood of a meteor strike in their future. We humans think we're much smarter and more adaptable than dinosaurs, and that because we have lots of sophisticated tools and big ideas about the past, present, and future, we can ride out whatever comes our way. That assumption may soon be put to the test.

If we want to minimize the impact of sea-level rise in the next century, here's how we do it: stop burning fossil fuels and move to higher ground. We wouldn't even have to stop burning fossil fuels tomorrow—if we did it by 2050, that would be good enough. It wouldn't entirely halt sea-level rise, but it would avoid the worst of it. Instead of six, seven, eight feet, or more by the end of the century, we might get two or three. We would still need to retreat from the low-lying coastlines, but instead of a stampede, it could be a leisurely stroll.

Unfortunately, it's hard to muster much faith that we will dramatically reduce CO_2 emissions anytime soon—much less cut them to zero by 2050. If that's the case, then planning for a world of rapidly rising seas will be much more difficult.

During my reporting for this book, I encountered a lot of thoughtful civic leaders and politicians who are thinking hard about how to reimagine the future in a world of fast-rising seas. In Norfolk, Virginia, city officials have collaborated with the US Navy and university researchers to come up with a comprehensive development plan for the year 2100 to help identify which neighborhoods are most at risk. The Southeast Florida Regional Climate Change Compact, which includes representatives of four counties in the region (including Miami-Dade), has pushed local and state officials to rethink zoning laws and cut bureaucratic impediments that delay aggressive measures to combat rising seas; the State of Louisiana released an ambitious but largely unfunded $50 billion master plan to save its sinking coastline and help protect New Orleans; in the UK, the government has encouraged a gentle retreat from the coasts through a "managed realignment" that encourages marshes and other coastal habitats to migrate inland, creating a natural buffer against the rising seas. In the Netherlands, they've been thinking about how to battle the sea for a thousand years and are now exporting that knowledge around the world. Wherever there is a city at risk of flooding, you'll likely find a Dutch engineer offering—or, just as often, selling—a solution.

These initiatives are all important, but they are just the preliminary sketches of the changes that need to be made in the decades ahead. The sheer economic chaos that looms for some coastal regions is hard to grasp, much less anticipate and prepare for. Nor do these initiatives begin to grapple with the political and psychological trauma of losing entire cities and coastlines, as well as the hopes and dreams that adhere to those cities and coastlines. As our planet changes, so will we.

Maybe in our increasingly rich and human-engineered world, losing some beaches and cities won't matter so much. If the real Venice goes under, you can always visit the fake Venice in Las Vegas. And maybe Miami Beach will be nearly as awesome in virtual reality. (Then again, maybe not.) Perhaps the best we can hope for is that living in a world of quickly rising seas will turn out to be a planetary-scale experiment in creative destruction, one that forces us to abandon a lot of stupid infrastructure and stupid ideas about how to live with water—and how to live with each other—and replace them with something smarter, more durable, more flexible. After all, other than cockroaches, humans are probably the most adaptable creatures on the planet. "To be honest, I'm looking forward to it," one Miami developer told me. "The only way we can survive this is to knock down a lot of old buildings and build new buildings that are higher, better, stronger. I joke with my wife, 'Hey, let's go buy a Hummer! Let's speed this thing up a little bit! How long do we have to wait?'"

The inundation of a modern coastal city is not something humans have ever witnessed before. We've seen floods and storms, but this will be nothing like what's to come. Even if it happens fast, it will seem to happen slowly. People are likely to notice in the same way they notice aging: *Oh, look how tall my kids are all of a sudden!*

In a similar way, people will notice higher tides that roll in more and more frequently. Water will pool longer in streets and parking lots. Trees will turn brown and die as they suck up salt water. Then a storm will hit, and it will push an astonishing amount of water into the city. Some people will move to new, higher buildings. Others will simply move to higher ground. Roads will be raised. Solar panels will bloom on rooftops. Abandoned houses will linger like ghosts, filling with feral cats and other refugees looking for their own high ground. Water will continue to creep in. It will have a metallic sheen and will smell bad. Kids will get strange rashes and fevers. More people will leave. Seawalls will crumble. In a few decades, low-lying neighborhoods will be knee-deep. Wooden houses will collapse into a sea of soda bottles, laundry detergent jugs, and plastic toothbrushes. Human bones, floated out of caskets, will be a common sight. Treasure hunters will kayak in, using small robotic submersibles to search for coins and jewelry. Modern office buildings and condo towers will lean as the salt water corrodes the concrete foundations and eats at the structural beams. Fish will school in classrooms. Oysters will grow on submerged light poles. Religious leaders will blame sinners for the drowning of the city. Journalists will arrive on floatplanes and write about the return of nature.

But mostly the city will be forgotten, one of many places lost to the attacking sea. In some distant future, someone, or some humanlike machines, may explore the sunken city and find bowling balls, stainless steel knives, gold wedding bands, and ceramic tiles. They may wonder about the people who lived there, what their lives were like, and what they were thinking as their world went under.

Acknowledgments

I'm grateful to everyone at Little, Brown for working so hard to launch this book into the world. Thanks to my editor, John Parsley, who believed in my work from the beginning, shaped it with patience and insight, and guided the book *almost* to the end. Thanks to Reagan Arthur for all her support and enthusiasm, to Michael Noon for pulling everything together, to Barbara Perris for her thoughtful copyediting, and to Gabriella Mongelli for her good-humored replies to even my crankiest emails. I'm grateful to Lauren Velasquez, Carrie Neill, and Elizabeth Garriga for making sure this book wasn't swamped by the media tides.

Heather Schroder, my agent of many years, understood even before I did that this was a book I needed to write. Every writer should be so lucky to have someone like Heather guiding their professional lives.

As a journalist, I'm also lucky to have had a home for the past two decades at *Rolling Stone*. I'm grateful to Jann

Wenner for many things, but especially for his unwavering belief that climate change is the biggest story of our time. For their editorial wisdom (and sweat), thanks to Jason Fine, Sean Woods, and Will Dana, who wrote the headline for the first story I ever published at the late, great Manhattan weekly *7 Days,* and who sent me on so many great journalistic adventures over the years. I'm grateful to Alison Weinflash for always being there when I needed her and to Elisabeth Garber-Paul and Coco McPherson for helping me keep my facts straight.

While writing this book, I spent two years as a Fellow at New America, where I benefited greatly from workshops and exchanges with other Fellows. For a journalist like me, being a Fellow at New America is like being adopted by an extended family where everyone works in the fields all day and feasts on big ideas at night. I'm especially grateful to Anne-Marie Slaughter, Kati Marton, and Peter Bergen for their support and friendship.

This book would not have been possible without the help of many scientists who gave me far more time and showed me far more patience than I had any right to expect. I'm deeply indebted to the late Peter Harlem, a Vietnam War veteran and dedicated scientist who shared his LiDAR elevation maps of South Florida with me and drove me around Miami in his MINI Cooper while narrating the story of sea-level rise, past, present, and future. Thanks to Hal Wanless for introducing me to Peter, for taking me into the mangroves, for all those meals at Burger Bob's in Coral Gables, for sharing his deep knowledge of Florida geology, and for having the courage to articulate what's at stake, even when it would have been much easier to keep quiet. I owe a special debt to

Andrea Dutton, who gave me an insightful tour of coral fossils in the Florida Keys and helped me with many questions about ancient ice and seas. Thanks to Jason Box for taking me to Greenland, where I saw not just melting ice sheets and calving glaciers, but a passionate scientist at work. Brian McNoldy, Keren Prize Bolter, Donald McNeill, Erin Lipp, Philip Orton, Jochen Hinkel, Ben Strauss, Richard Kopp, Richard Alley, Peter Clark, Rob DeConto, and James Hansen were all generous with their time.

Many people helped me in big and small ways during the three years I worked on this book. It's impossible to name them all, but I would like to single out Sheryl Gold, Roni Avissar, Philip Stoddard, Albert Slap, Wayne Pathman, David Martin, Bruce Mowry, Alastair Gordon, Barbara De Vries, John Stuart, Richard Saltrick, Michael Gerrard, Karen O'Neil, Susannah Drake, Manuel Rosa da Silva, Dan Zarrilli, Nancy Kete, Jo da Silva, Henk Ovink, Richard Jorissen, Miranda Mens, Peter Persoon, Thom Woodroofe, Dean Bialek, Megan Chapman, Andrew Maki, Brian Deese, Jon Finer, and David Keith. And a special thanks to Brian Palmateer, Pat Palmateer, and Jeff Kelleher for the music of hammers and saws that inspired me to keep typing when I really wanted to go to the beach.

I owe an especially big debt to Reinaldo Borges for sharing his infectious love of Miami; to Dan Dudek for his friendship and his wisdom about climate policy; to Kevin Knobloch and Nicole St. Clair Knobloch for their wit and whiskey; to Aimee Christensen for her political savvy and indefatigable spirit; and to Jim and Karen Shepard for reminding me, at a particularly difficult moment, that writing is supposed to be fun.

This book is dedicated to Milo, Georgia, and Grace, who proved yet again that putting up with a distracted and work-worn father is no barrier to growing up to become remarkable human beings.

Finally, I owe a great deal more than a thank-you to Pernille. While I was writing this book, she rode all the big waves with me and helped me avoid the sea monsters below. I am grateful for the countless lunches, her editorial insight, the long walks around Lisbon and Copenhagen; and her appreciation of the beauty of the world we live in — and the beauty of the world that is yet to come.

Notes

Prologue

Frost: Interview with Phillip Frost. *The Sunshine State.* WLRN, June 12, 2017.

Hansen: Jeff Goodell. "We Must Act Now." *Rolling Stone*, August 20, 2009, 65.

twenty to thirty feet: Andrea Dutton et al. "Sea-Level Rise Due to Polar Ice-Sheet Mass Loss During Past Warm Periods." *Science* 349, no. 6244 (2015).

four hundred feet: Richard B. Alley et al. "Ice-Sheet and Sea-Level Changes." *Science* 310, no. 5747 (October 21, 2005), 457.

six inches: Robert Kopp et al. "Temperature-Driven Global Sea-Level Variability in the Common Era." *Proceedings of the National Academy of Sciences* 113, no. 11 (2016), 1435.

twice the rate: Carling C. Hay et al. "Probabilistic Reanalysis of Twentieth-Century Sea-Level Rise." *Nature* 517, no. 7535 (2015), 481.

eight feet by 2100: The estimated sea-level rise for 2100 in the most recent report by the Intergovernmental Panel on Climate Change is twenty-six to ninety-eight centimeters (about one foot to three feet). But this does not include contributions from marine-based ice sheets in Antarctica, in part because, at the time the IPCC report was finalized, there was not enough confidence in scientists' understanding of the dynamics of these ice sheets to make any sound projections. (New research published since the IPCC report was finalized has resolved some of that uncertainty.) See John Church and Peter Clark et al. *Climate Change 2013: The Physical Science Basis. Contribution of Working Group I to the Fifth Assessment Report of the Intergovernmental Panel on Climate Change* (Cambridge and New York: Cambridge University Press, 2013). https://www.ipcc.ch/report/ar5/wg1/

In 2017, the US's National Oceanic Atmospheric Administration did its own evaluation of future sea-level rise, which included more recent papers on ice dynamics on main Antarctica. Not surprisingly, the NOAA paper comes up with bigger numbers than did the IPCC, suggesting that we could see between 30 centimeters and 2.5 meters of sea-level rise by 2100 (one foot to more than eight feet). See William Sweet et al. "Global and Regional Sea-Level Rise Scenarios for the United States." NOAA technical report January 2017, vi. https://tidesandcurrents.noaa.gov/publications/techrpt83_Global_and_Regional_SLR_Scenarios_for_the_US_final.pdf

hottest year: "NASA, NOAA Data Show 2016 Warmest Year on Record Globally." NASA press release, January 18, 2017. Accessed March 3, 2017. https://www.nasa.gov/press-release/nasa-noaa-data-show-2016-warmest-year-on-record-globally

thirty-six degrees: Chris Mooney and Jason Samenow. "The North Pole Is an Insane Thirty-Six Degrees Warmer Than Normal as Winter Descends." *Washington Post,* November 17, 2016.

thirteen feet: Pierre Deschamps et al. "Ice-Sheet Collapse and Sea-Level Rise at the Bølling Warming 14,600 Years Ago." *Nature* 483, no. 7391 (2012), 559.

"Stonehenge": David Archer. *The Long Thaw: How Humans Are Changing the Next 100,000 Years of Earth's Climate* (Princeton: Princeton University Press, 2008), 1.

time to adapt: Sweet et al., "Global and Regional Sea-Level Rise Scenarios for the United States," 12.

two hundred feet: Ricarda Winkelmann et al. "Combustion of Available Fossil Fuel Resources Sufficient to Eliminate the Antarctic Ice Sheet." *Science Advances* September 11, 2015, vol. 1, no. 8.

Zillow: Krishna Rao. "Climate Change and Housing: Will a Rising Tide Sink All Homes?" Zillow, August 2, 2016. Accessed March 2, 2017. https://www.zillow.com/research/climate-change-underwater-homes-12890/

$100 trillion: Jochen Hinkel et al. "Coastal Flood Damage and Adaptation Costs under 21st Century Sea-Level Rise." *Proceedings of the National Academy of Sciences* 111, no. 9 (2014), 3292.

145 million people: David Anthoff et al. "Global and Regional Exposure to Large Rises in Sea-Level: A Sensitivity Analysis." Report by the Tyndall Centre for Climate Change Research (2006), 8.

people will be displaced: Benjamin H. Strauss et al. "Mapping Choices: Carbon, Climate, and Rising Seas, Our Global Legacy." Climate Central Research Report (November 2015), 5. A slightly different way of looking at populations of displaced people can be found in Robert Nicholls et al. "Sea-level rise and Its Possible Impacts Given a 'Beyond 4 C World' in the Twenty-first Century." *Philosophical Transactions of the Royal Society* 369 (2011), 161–181. A good discussion of the complexity of estimating the number of people who will be displaced by rising seas can be found in Michal Lichter et al. "Exploring Data-Related Uncertainties in Analyses of Land Area and Population in the 'Low-Elevation Coastal Zone.'" *Journal of Coastal Research,* vol. 27, no. 4 (July 2011), 757–768.

Chapter 1

R/V *Knorr:* Kathryn Eident. "Farewell to the *Knorr.*" *Oceanus,* December 1, 2014.

14,500 years ago: Jessi Halligan et al. "Pre-Clovis Occupation 14,550 Years Ago at the Page-Ladson Site, Florida, and the Peopling of the Americas." *Science Advances* 2, no. 5 (May 1, 2016), e1600375.

more than a foot per decade: Pierre Deschamps et al. "Ice-Sheet Collapse and Sea-Level Rise at the Bølling Warming 14,600 Years Ago." *Nature* 483, no. 7391 (March 29, 2012), 559.

five hundred to six hundred feet a year: Personal communication with Jessi Halligan, October 2016.

A Grammar of Yidiɲ: R. M. W. Dixon, *A Grammar of Yidiɲ* (Cambridge: Cambridge University Press, 1972).

"In the beginning, as far back as we remember...": Patrick D. Nunn and Nicholas J. Reid. "Aboriginal Memories of Inundation of the Australian Coast Dating from More Than 7,000 Years Ago." *Australian Geographer* 47, no. 1 (September 7, 2015), 12.

islands like Fitzroy: Ibid., 26.

even earlier flood story in *The Epic of Gilgamesh:* Irving Finkel. *The Ark Before Noah* (New York: Anchor, 2014), 35.

"It's hard to imagine the terror...": William Ryan and Walter Pitman. *Noah's Flood* (New York: Simon and Schuster, 2000), 235.

It's not a thesis all scientists accept: Liviu Giosan et al. "Was the Black Sea Catastrophically Flooded in the Early Holocene?" *Quaternary Science Reviews* 28, no. 1 (January 2009), 1–6.

Calusa: Victor D. Thompson et al. "From Shell Midden to Midden-Mound:

The Geoarchaeology of Mound Key, an Anthropogenic Island in Southwest Florida, USA." Karen Hardy, ed. *PLoS ONE* 11, no. 4 (April 28, 2016), 46.

Chapter 2

$25-million Mediterranean Revival mansion: Debora Lima. "Former Miami Beach Home of Lenny Kravitz Listing for $25 Million." *Miami Herald,* March 27, 2016.

"each of us began": Rachel Carson. *The Sea Around Us* (New York: Oxford University Press, 1951), 14.

A few billion years ago: Albert C. Hine. *Geologic History of Florida* (Gainesville: University Press of Florida, 2013), 47.

ooids: Ibid., 197.

In the pancake-flat topography: Ibid., 199.

Julia Tuttle: T. D. Allman. *Finding Florida* (New York: Grove/Atlantic, 2013), 320.

the memoirs of George Merrick: Arva Moore Parks. *George Merrick, Son of the South Wind* (Gainesville: University Press of Florida, 2015), 68–72.

By 1909, dredging of the Miami Canal: T. D. Allman. *Miami* (Gainesville: University Press of Florida, 2013), 239.

"From the time the Hebrews...": Michael Grunwald. *The Swamp: The Everglades, Florida, and the Politics of Paradise* (New York: Simon and Schuster, 2007), 176.

"The suntan, once a symbol of labor...": Ibid.

"There is something very distressing...": Ibid., 174.

"Virgin jungle crept right down...": Jerry M. Fisher. *The Pacesetter* (Victoria, BC: FriesenPress, 2014), 208.

Jane Austen observed: Cited in John R. Gillis. *The Human Shore* (Chicago: University of Chicago Press, 2012), 115.

"aesthetic conquest of the shore...": Ibid., 149.

Jack Peacock: Fisher, *The Pacesetter,* 141.

a New Jersey farmer named John Collins: Ibid., 142.

"Creatures that made me shudder": Mark Davis, ed. *American Experience:* "Mr. Miami Beach." WGBH, Boston, 1998.

"sloppy Cream of Wheat": Fisher, *The Pacesetter,* 161.

"The bayside gradually began": Ibid., 165.

"Hundreds of Negroes": Davis, *American Experience:* "Mr. Miami Beach."

Venice of America: Beth Duff Sanders. "Affluent Area Has Problems and Squabbles Too." *Sun Sentinel,* April 26, 1989.

"Hardly anybody talks of anything…": Davis, *American Experience:* "Mr. Miami Beach."

"the first man smart enough to discover…": Fisher, *The Pacesetter,* 320.

"Miami Beach was isolated in a sea…": Quoted in Polly Redford. *Billion-Dollar Sandbar: A Biography of Miami Beach* (New York: Dutton, 1970), 123.

September 18, 1926: Fisher, *The Pacesetter,* 300–304.

fifteen-foot tsunami: Grunwald, *The Swamp,* 192.

"Miami Beach has based its economy…": Jerry Iannelli. "Miami Beach Plans to Use Alarming Ads to Scare Away Airbnb-Style Renters." *Miami New Times,* September 8, 2016.

"the poor people who suffered…": Grunwald, *The Swamp,* 180.

"Sure, some lives were lost…": Ibid., 188.

Chapter 3

entire surface of the ice sheet: "An Intense Greenland Melt Season: 2012 in Review." Nsidc.org, February 5, 2013. Accessed February 12, 2017. http://nsidc.org/greenland-today/2013/02/greenland-melting-2012-in-review/

River Thames: According to Jason Box, peak discharge on the Watson River during the 2012 heat wave averaged 1,200 cubic meters per second and peaked at 3,200 cubic meters per second. This highest recorded flow on the Thames was 275 cubic meters per second in January 2014. See Matt McGrath. "River Thames Breaks Records for Water Flows in January." *BBC News,* February 13, 2014. Accessed May 3, 2017. http://www.bbc.com/news/science-environment-26175213

risks in Greenland and Antarctica: "Quick Facts on Ice Sheets." National Snow and Ice Data Center. Nsidc.org. Accessed February 19, 2017. https://nsidc.org/cryosphere/quickfacts/icesheets.html

seven billion human beings: "Richard Alley at INSTAAR, April 2015." University of Colorado Boulder, April 6, 2015.

West Antarctica: John Mercer. "West Antarctic Ice Sheet and CO_2 Greenhouse Effect: A Threat of Disaster." *Nature* 271, 1978, 1–5. See also Ian Joughin et al. "Marine Ice Sheet Collapse Potentially Under Way for the Thwaites Glacier Basin, West Antarctica." *Science* 344 (May 2014), 735–38.

raise the seas ten to thirteen feet: Personal communication with Penn State glaciologist Richard Alley, February 7, 2017.

glaciologist Jay Zwally: Jay Zwally et al. "Surface Melt–Induced Acceleration of Greenland Ice-Sheet Flow." *Science* 297, no. 5579 (July 12, 2002), 218–22.

fingerprinting: Jerry X. Mitrovica et al. "The Sea-Level Fingerprint of West Antarctic Collapse." *Science* 323, no. 5915 (February 6, 2009), 753–53.

water would rise thirteen feet: William Sweet et al. "Global and Regional Sea Level Rise Scenarios for the United States." *NOAA Technical Report* NOS CO-OPS 083 (January 2017), 17.

three times as much ice: Darryl Fears. "New Study Affirms Ice-Sheet-Loss Estimates in Greenland, Antarctica." *Washington Post,* November 29, 2012.

trillion tons: Chelsea Harvey. "Greenland Lost a Staggering One Trillion Tons of Ice in Just Four Years." *Washington Post,* July 19, 2016.

Crystal Serenity: Will Oremus. "The Upside of Global Warming: Luxury 'Northwest Passage' Cruises for the Filthy Rich." Slate, August 17, 2016. Accessed March 3, 2017. http://www.slate.com/blogs/future _tense/2016/08/17/crystal_serenity_s_northwest_passage _cruise_is_a_festival_of_environmental.html

three degrees Fahrenheit: Andrea Thompson. "2016 'Arctic Report Card' Gives Grim Evaluation." Climate Central, December 14, 2016. Accessed February 20, 2017. http://www.climatecentral.org/ news/2016-arctic-report-card-grim-20968

California's recent record drought: S. Y. Wang et al. "Probable Causes of the Abnormal Ridge Accompanying the 2013–2014 California Drought: ENSO Precursor and Anthropogenic Warming Footprint." *Geophysical Research Letters* 41, no. 9 (May 16, 2014), 3220–26.

summer climate extremes: Michael Mann et al. "Influence of Anthropogenic Climate Change on Planetary Wave Resonance and Extreme Weather Events." *Scientific Reports* 7 (March 27, 2017), 45242.

"Scientists didn't expect to see…": Author interview with Michael Mann, June 2013.

giant cloud of water: "Astronomers Find Largest, Most Distant Reservoir of Water." NASA.gov, July 22, 2011. https://www.nasa.gov/topics/ universe/features/universe20110722.html

scientists have calculated: N. B. Karlsson et al. "Volume of Martian Mid-latitude Glaciers from Radar Observations and Ice Flow Modeling." *Geophysical Research Letters,* 42 (April 28, 2015), 2627.

three feet of ice: Cassie Stuurman et al. "SHARAD detection and characterization of subsurface water ice deposits in Utopia Planitia, Mars." *Geophysical Research Letters* 43 (2016), 9484–9491.

icy comets: Brian Greene. "How Did Water Come to Earth?" *Smithsonian,* May 2013.

Milutin Milankovitch: Steve Graham. "Milutin Milankovitch." NASA Earth Observatory, March 24, 2000. Accessed March 1, 2017. https://earthobservatory.nasa.gov/Features/Milankovitch/

Thomas Jefferson: Amy Dusto. "Reading Between the Tides: 200 Years of Measuring Global Sea Level." Climate.gov, August 4, 2014. Accessed February 20, 2017. https://www.climate.gov/news-features/climate -tech/reading-between-tides-200-years-measuring-global-sea -level

glacial rebound: "Glacial Rebound: The Not So Solid Earth." NASA.gov, August 26, 2015. Accessed March 10, 2017. http://www.nasa.gov/ feature/goddard/glacial-rebound-the-not-so-solid-earth

thermal expansion: John Church and Peter Clark et al. *Climate Change 2013: The Physical Science Basis. Contribution of Working Group I to the Fifth Assessment Report of the Intergovernmental Panel on Climate Change,* 1161.

Cape Hatteras: Paul B. Goddard et al. "An Extreme Event of Sea-Level Rise Along the Northeast Coast of North America in 2009–2010." *Nature Communications* 6 (February 24, 2015), 6346.

time machine: Jerry X. Mitrovica et al. "Reconciling Past Changes in Earth's Rotation with Twentieth-Century Global Sea-Level Rise: Resolving Munk's Enigma." *Science Advances* 1, no. 11 (December 1, 2015), e1500679–79.

"Jason was smart…": Cited in Jeff Goodell. "The Ice Maverick." *Rolling Stone,* August 3, 2013.

James Hansen: James Hansen and Larissa Nazarenko. "Soot Climate Forcing via Snow and Ice Albedos." *Proceedings of the National Academy of Sciences* 101, no. 2 (January 13, 2004), 423–28.

collapsed in spectacular fashion: "Larsen B Ice Shelf Collapses in Antarctica." Nsidc.org, March 18, 2002. Accessed March 2, 2017. https://nsidc.org/news/newsroom/larsen_B/2002.html

"It was mind-blowing…": Interview with the author, July 2016.

"It showed us how much…": Interview with the author, August 2016.

2013 IPCC report: John Church and Peter Clark et al. *Climate Change 2013: The Physical Science Basis. Contribution of Working Group I to the Fifth Assessment Report of the Intergovernmental Panel on Climate Change.*

exponential increase: James Hansen et al. "Ice Melt, Sea-Level Rise, and Superstorms: Evidence from Paleoclimate, Data, Climate Modeling and Modern Observations that 2°C Global Warming Could Be Dangerous." *Atmospheric Chemistry and Physics Discussions* 23 (2015), 20063.

three feet of sea-level rise by 2100: Robert DeConto and David Pollard. "Contribution of Antarctica to Past and Future Sea-Level Rise." *Nature* 531, no. 7596 (March 30, 2016), 591.

"OMG thing…": Quoted in Don Jergler. "RIMS 2016: Sea-Level Rise Will Be Worse and Come Sooner." *Insurance Journal,* April 12, 2016, 64.

Chapter 4

$100 million: Erica Martinson. "Obama's Budget Shows Alaska's on the President's Mind." *Alaska Dispatch News,* February 9, 2016.

twice as fast: "Climate Impacts in Alaska." EPA.gov. Accessed February 21, 2017. https://www.epa.gov/climate-impacts/climate-impacts-alaska

thirty-five thousand stressed-out walruses: The Associated Press. "Alaska: Walrus Again Crowd onto Shore." *New York Times,* September 10, 2015.

$3.7-billion budget shortfall: Tim Bradner. "Fiscal Year 2016 Budget Deficit Estimated at $3.7 Billion." *Alaska Journal of Commerce,* July 8, 2015.

30 percent of the known natural gas reserves: US Energy Information Administration. "Arctic Oil and Natural Gas Resources." Eia.gov, January 20, 2012. Accessed March 4, 2017. http://www.eia.gov/todayinenergy/detail.php?id=4650

sixty feet of shoreline: Nichelle Smith and Sattineni Anoop. "Effect of Erosion in Alaskan Coastal Villages." Proceedings of Fifty-Second ASC Annual International Conference, 2016, 98.

Chapter 5

"the Trump of the Tropics": Siobhan Morrissey. "Twenty-Five Most Influential Hispanics in America." *Time,* August 22, 2005.

$55 million: "Museum Receives $40 Million Gift from Miami Developer Jorge M. Pérez." PAMM.org, December 1, 2011. Accessed March 2, 2017. http://www.pamm.org/about/news/2011/museum-receives-40-million-gift-miami-developer-jorge-m-pérez

See also "Pérez Art Museum Miami Receives $15 Million Gift from Philanthropist and Patron of the Arts Jorge M. Pérez." PAMM.org, November 30, 2016. Accessed March 2, 2017. http://pamm.org/about/news/2016/pérez-art-museum-miami-receives-15-million-gift-philanthropist-and-patron-arts-jorge

one out of every five condos: Personal communication with condo analyst Peter Zalewski, February 21, 2017.

"He dares to dream…": Jorge Pérez. *Powerhouse Principles* (New York: Penguin, 2008), xi.

$2.8 billion: "The Forbes 400." Forbes.com. Accessed February 21, 2017. http://www.forbes.com/profile/jorge-perez/

More than three quarters of the population: "National Coastal Population Report." NOAA's State of the Coast website, March 2013.

the Risky Business Project: "Risky Business: The Economic Risks of Climate Change in the United States." The Risky Business Project, June 2014.

one third: Personal communication with Miami-Dade Tax Collector's Office, January 20, 2017.

Foreign nationals: Nichola Nehamas. "Buying a Home in Miami-Dade Is So Expensive It Could Hurt the Economy." *Miami Herald,* February 9, 2017.

Cash deals: "Feds Want to Know Who's Behind Purchases in Number One Cash Real Estate Market Miami." Zillow, January 13, 2016. Accessed March 2, 2017. https://www.zillow.com/blog/cash-buyers-in-real -estate-market-190774/

$25 billion: "Hurricanes in History." National Hurricane Center. Accessed February 21, 2017. http://www.nhc.noaa.gov/outreach/ history/#andrew

$2 billion in additional damage: Harvey Liefert. "Sea-Level Rise Added $2 Billion to Sandy's Toll in New York City." *Eos,* March 16, 2015. Accessed May 1, 2017. https://eos.org/articles/sea-level-rise-added -2-billion-to-sandys-toll-in-new-york-city

$23 billion in debt: United States Government Accountability Office. "GAO Report to Congressional Committees: High Risk Series." February 2017.

$428 billion in property value: See "Policy Information by State" section at FEMA.gov. Accessed May 5, 2017. https://bsa.nfipstat.fema.gov/ reports/1011.htm

346,742 policies: Ibid.

modest reforms: Ann Carrns. "Federal Flood Insurance Premiums for Homeowners Rise." *New York Times,* April 2, 2015.

ten thousand properties: Jake Martin. "Proposed FEMA Maps Remove over 10,000 Structures from St. Johns County Flood Zones." *St. Augustine Record,* July 14, 2016.

Royal Russian Midgets: Theo Karantsalis. "Sweetwater's History Rich with Circus-Like Troubles." *Miami Herald,* December 12, 2014.

poorest cities: "QuickFacts: Sweetwater, Florida." *United States Census Bureau.* Accessed February 21, 2017. www.census.gov/quickfacts/ table/PST045216/1270345,12,56037,00

"ground zero": Marc Caputo. "2013: A Dirty Year When It Came to Public Corruption in Miami-Dade." *Miami Herald,* December 28, 2013.

$17-million budget: "City of Sweetwater Adopted Budget FY 2016–2017." Cityofsweetwater.fl.gov, 2017. Accessed March 1, 2017. http://city ofsweetwater.fl.gov/documents/Budget%202016-2017%20 FINAL%20ADOPTED%20BUDGET.pdf

Chapter 6

"Time is water...": Charles Fenyvesi. "The City Nobel Laureate Joseph Brodsky Called Paradise." *Smithsonian Journeys,* Winter, 2015, 68–72.

Giovanni da Cipelli: Egnazio's edict, translated from engraving on a marble slab in Museo Correr, Venice. http://correr.visitmuve.it/ en/il-museo/layout-and-collections/venetian-culture/

"islands were submerged...": Thomas F. Madden. *Venice: A New History* (New York: Viking, 2012), 63.

Marco Polo's house: John Berendt. *The City of Falling Angels* (New York: Penguin, 2006), 183.

"difficult to overestimate": Madden, *Venice,* 412–413.

Special Law of 1973: Ibid., 412.

corruption scandal: "Mayor of Venice Arrested over Alleged Bribes Relating to Flood Barrier Project." *The Guardian,* June 4, 2014.

Aquagranda: Opera composed by Filippo Perocco, libretto by Roberto Bianchin and Luigi Cerantola. Premiered November 4, 2016, at La Fenice, Venice.

five hundred police officers: Nick Squires. "Venice Dawn Raids over Flood Barrier Corruption." *The Telegraph,* July 12, 2013.

"illicit gains": Cited in Salvatore Settis. *If Venice Dies* (New York: New Vessel Press, 2016), 171.

UNESCO report: "From Global to Regional: Local Sea-Level Rise Scenarios." Report of workshop organized by UNESCO Venice office, November 22–23, 2010.

1953 flood: Tracy Metz and Maartje van den Heuvel. *Sweet and Salt* (Rotterdam: NAi Publishers, 2012), 227.

"O Venice!": From "Ode to Venice." Collected in George Gordon Byron. *Lord Byron: The Major Works* (London: Oxford University Press, 2008), 301.

Chapter 7

Hurricane Sandy: Storm damage statistics from personal communication with Office of Mayor Bill de Blasio.

East Side Coastal Resiliency Project: "OneNYC: 2016 Progress Report." The City of New York, Mayor Bill de Blasio. May 2016, 160.

budgeted at $760 million: Personal communication with Dan Zarrilli.

10 percent of the US gross domestic product: Richard Florida. "Sorry, London: New York Is the World's Most Economically Powerful City." TheAtlantic.com, March 3, 2015. Accessed March 1, 2017. http://www.citylab.com/work/2015/03/sorry-london-new-york-is-the-worlds-most-economically-powerful-city/386315/

Manhattan in 1650: Snejana Farberov. "How Hurricane Sandy Flooded New York Back to Its Seventeenth-Century Shape as It Inundated 400 Years of Reclaimed Land." *Daily Mail,* June 16, 2013.

$129 billion: "On the Front Lines: $129 Billion in Property at Risk from Floodwaters." Office of the New York City Comptroller, October 2014, 2.

50 percent faster: Robert Kopp et al. "Probabilistic Twenty-First and Twenty-Second-Century Sea-Level Projections at a Global Network of Tide-Gauge Sites." *Earth's Future* 2, no. 8 (August 1, 2014), 383–406.

Rebuild by Design competition: Rebuild by Design website. Accessed March 2, 2017. http://www.rebuildbydesign.org/our-work/exhibitions/rebuild-by-design-hurricane-sandy-design-competition-exhibition. See also Rory Stott. "OMA and BIG Among Six Winners in Rebuild by Design Competition." ArchDaily, June 3, 2014. Accessed March 4, 2017. http://www.archdaily.com/512516/oma-wins-rebuild-by-design-competition-with-resist-delay-store-discharge

Sandy-like events: A. J. Reed et al. "Past, Present, and Future Threat of Tropical Cyclones and Coastal Flooding in New York City." American Geophysical Union fall meeting abstracts, December 1, 2015.

Bangladesh: Quirin Schiermeier. "Floods: Holding Back the Tide." *Nature* 508 (April 10, 2014), 164–166.

seawalls in Zhuhai: Michael Kimmelman. "Rising Waters Threaten China's Rising Cities." *New York Times,* April 7, 2017.

Susannah Drake: An overview of Drake's plan for Lower Manhattan is available on her firm's website. Accessed May 5, 2017. http://www.dlandstudio.com/projects_moma.html

Living Breakwaters: SCAPE's plan for Staten Island is available on the website of the New York State Governor's Office of Storm Recovery. Accessed May 5, 2017. https://stormrecovery.ny.gov/living-breakwaters-tottenville

Orff wrote: Kate Orff. "Adapt to the Future with Landscape Design." *New York Times,* October 28, 2015.

Blue Dunes: An overview of Blue Dunes is available on the West 8 website. Accessed May 5, 2017. http://www.west8.nl/projects/all/blue_dunes_the_future_of_coastal_protection/

high-level study: Cynthia Rosenzweig et al. *Responding to Climate Change in New York State: The ClimAID Integrated Assessment for Effective Climate Change Adaptation.* Technical report. New York State Energy Research and Development Authority (NYSERDA), Albany.

LaGuardia Airport: Deepti Hajela. "New York Reveals $4 Billion Plan for a New LaGuardia Airport." The Associated Press, July 27, 2015.

case study: Mireya Navarro. "New York Is Lagging as Seas and Risks Rise, Critics Warn." *New York Times,* September 10, 2012.

guidelines: Nicholas Kusnetz. "NYC Creates Climate Change Roadmap for Builders: Plan for Rising Seas." *InsideClimate News,* May 3, 2017. Accessed May 8, 2017. https://insideclimatenews.org/news/0205 2017/nyc-publishes-building-design-guidelines-adapting -climate-change

Broad Channel: Lisa L. Colangelo. "Queens Residents Still Struggle to Rebuild Homes Damaged by Hurricane Sandy Two Years Ago." *New York Daily News,* October 26, 2014.

Army Corps of Engineers: "Atlantic Coast of New York, East Rockaway Inlet to Rockaway Inlet and Jamaica Bay." Report by US Army Corps of Engineers New York District, August 2016.

Chapter 8

"Within seconds...": Corel Davenport. "The Marshall Islands Are Disappearing." *New York Times,* December 1, 2015.

twenty-three bombs: "Bikini Atoll Nuclear Test Site." UNESCO World Heritage List. Accessed March 1, 2017. http://whc.unesco.org/en/list/1339

the US argued: "The Legacy of US Nuclear Testing and Radiation Exposure in the Marshall Islands." Report by the US Embassy in the Republic of the Marshall Islands. Accessed March 7, 2017. https://mh.usembassy.gov/the-legacy-of-u-s-nuclear-testing-and -radiation-exposure-in-the-marshall-islands/

"Within hours...": Oliver Milman and Mae Ryan. "In the Marshall Islands, Climate Change Knocks on the Front Door." *Newsweek,* September 15, 2016.

Cancers: Steven L. Simon et al. "Radiation Doses and Cancer Risks in the Marshall Islands Associated with Exposure to Radioactive Fallout from Bikini and Enewetak Nuclear Weapons Tests: Summary." *Health Physics* 99, no. 2 (August 1, 2010), 105.

"national hero": Lisa Friedman. "Tony de Brum, Voice of Pacific Islands on Climate Change, Dies at 72." *New York Times,* August 22, 2017.

Portland, Oregon: This is a rough estimate. The Marshall Islands emitted about 125,000 metric tons of CO_2 equivalent in 2000. See

"Republic of the Marshall Islands Intended Nationally Determined Contribution." Report to the UNFCC, July 21, 2015. According to Kyle Diesner, policy analyst for the city of Portland, total emissions in Multnomah County, Oregon, in 2014 were 7,064,000 metric tons of CO_2 equivalent. Here's the math: 125,000 tons x 50 years = 6,250,000 tons < 7,064,000.

"genocide": "Climate Change Migration Is Cultural Genocide." Tony de Brum interview on Radio New Zealand, October 6, 2015. Accessed March 1, 2017. http://www.radionz.co.nz/international/programmes /datelinepacific/audio/201773361/climate-change-migration -is-cultural-genocide-tony-de-brum

unemployment rate: *The World Factbook: 2013–14* (Washington, DC: Central Intelligence Agency, 2013).

twelve hundred Americans: "Reagan Test Site, Marshall Islands: Managing a Missile Test Range Crucial to US Defense." Bechtel. Accessed January 14, 2017. http://www.bechtel.com/projects/kwajalein-test -range/

new radar installation: Nick Perry. "US Ignored Rising Sea Warnings at Radar Site." The Associated Press, October 18, 2016.

34 million gallons: "Fresh Water Sources." Marshall Islands Guide. October 9, 2015. Accessed January 20, 2017. http://www.infomarshallis lands.com/fresh-water-sources/

overweight or obese: W. Snowdon and A. M. Thow. "Trade Policy and Obesity Prevention: Challenges and Innovation in the Pacific Islands." *Obesity Reviews* 14, no. 2 (October 23, 2013), 150–58.

In Egypt: Jean-Daniel Stanley and Pablo L. Clemente. "Increased Land Subsidence and Sea-Level Rise Are Submerging Egypt's Nile Delta Coastal Margin." *GSA Today* 27, no. 5. (May 2017).

World Bank: Susmita Dasgupta et al. "River Salinity and Climate Change: Evidence from Coastal Bangladesh." World Bank Group, Policy Research working paper, March 2014.

shrimp farming: Joanna Lovatt. "The Bangladesh Shrimp Farmers Facing Life on the Edge." *The Guardian,* February 17, 2016.

San Diego: David Talbot. "Desalinization out of Desperation." *MIT Technology Review,* December 16, 2014.

low-elevation coastal zones: Generally defined as thirty feet or less below high-tide line. See Barbara Neumann et al. "Future Coastal Population Growth and Exposure to Sea-Level Rise and Coastal Flooding—A Global Assessment." *PLoS ONE* vol. 10, issue 3 (2015).

Green Climate Fund: Michael Slezak. "Obama Transfers $500 Million to Green Climate Fund in Attempt to Protect Paris Deal." *The Guardian,* January 17, 2017.

"a Chinese colony": Sanjay Kumar. "This Will Make the Country a Chinese Colony." TheDiplomat.com, July 25, 2015. Accessed May 4, 2017. http://thediplomat.com/2015/07/this-will-make-the-country-a-chinese-colony/

200 million climate refugees: "Migration, Climate Change, and the Environment: A Complex Nexus." International Organization for Migration website. Accessed January 24, 2017. https://www.iom.int/complex-nexus#estimates

1951 Refugee Convention: "What Is a Refugee?" The UN Refugee Agency website. Accessed March 2, 2017. http://www.unrefugees.org/what-is-a-refugee/

fifty chickens a minute: Bryce Covert. "The Hellish Conditions Facing Workers at Chicken Processing Plants." Thinkprogress.com, October 27, 2015. Accessed May 5, 2017. https://thinkprogress.org/the-hellish-conditions-facing-workers-at-chicken-processing-plants-1eb2f4206968

Human Rights Watch: "Blood, Sweat, and Fear: Workers' Rights in U.S. Meat and Poultry Plants." Report by Human Rights Watch (January 24, 2005), 32.

Kiribati purchased: Laurence Caramel. "Besieged by the Rising Tides of Climate Change, Kiribati Buys Land in Fiji." *The Guardian,* June 30, 2014.

"Fiji will not turn...": Shalveen Chand. "Kiribati's Hope for Land." *Fiji Times,* February 14, 2014.

"You can drastically...": Quoted in Michael Gerrard. "America Is the Worst Polluter in the History of the World. We Should Let Climate Change Refugees Resettle Here." *Washington Post,* June 25, 2015.

Gerrard argues: Ibid.

Runit Dome: Michael Gerrard. "A Pacific Isle, Radioactive and Forgotten." *New York Times,* December 3, 2014.

Chapter 9

$250 million: "On the Front Lines of Rising Seas: Naval Station Norfolk, Virginia." Fact sheet, Union of Concerned Scientists. Accessed March 20, 2017. http://www.ucsusa.org/global-warming/global-warming-impacts/sea-level-rise-flooding-naval-station-norfolk#.WMqvUBiZNN0

political instability: Benjamin I. Cook et al. "Spatiotemporal Drought Variability in the Mediterranean over the Last 900 Years." *Journal of Geophysical Research: Atmospheres* vol. 121, no. 5 (2016), 2060–2074.

real estate portfolio: Chuck Hagel. Quadrennial Defense Review. US Department of Defense, March 4, 2014, 40.

Diego Garcia: "Military Expert Panel Report: Sea-Level Rise and the US Military's Mission." The Center for Climate and Security, September 2016, 67.

witch-hunt: John Collins Rudolf. "A Climate Skeptic with a Bully Pulpit in Virginia Finds an Ear in Congress." *New York Times,* February 22, 2011.

"left-wing term": Rebecca Leber. "Virginia Lawmaker Says 'Sea-Level Rise' Is a 'Left-Wing Term,' Excises It from State Report on Coastal Flooding." ThinkProgress, June 10, 2012. Accessed March 3, 2017. https://thinkprogress.org/virginia-lawmaker-says-sea-level-rise-is-a-left-wing-term-excises-it-from-state-report-on-coastal-805134396adc?gi=7f60ed42a9be#.3xm2pbnlv

Governor Terry McAuliffe: Ladelle McWhorter and Mike Tidwell. "Virginia Governor Terry McAuliffe Has Abysmal Climate Record." *Washington Post,* June 10, 2016.

One study: "Recurrent Flooding Study for Tidewater Virginia." Virginia Institute of Marine Sciences, January 2013.

Arab Spring: Francesco Femia and Caitlin Werrell, eds. "The Arab Spring and Climate Change." The Center for Climate and Security, February 2013.

Boko Haram: Erika Eichelberger. "How Environmental Disaster Is Making Boko Haram Violence Worse." MotherJones.com, June 10, 2014. Accessed May 3, 2017. http://www.motherjones.com/environment/2014/06/nigeria-environment-climate-change-boko-haram

Rear Admiral Daniel Abel: Emily Russo Miller. "For New Coast Guard Head, Mission Still the Same." *Juneau Empire,* December 7, 2014.

six Russian fighters: Steve Brusk and Ralph Ellis. "Russian Planes Intercepted near US, Canadian Airspace." CNN.com, November 13, 2014. Accessed March 7, 2017. http://www.cnn.com/2014/09/19/us/russian-plane-incidents/

Bulava: Trude Pettersen. "One More Missile Launch from Barents Sea." *Barents Observer,* November 5, 2014.

"fifteenth century...": Doug Struck. "Russia's Deep-Sea Flag-Planting at North Pole Strikes a Chill in Canada." *Washington Post,* August 7, 2007.

2011 op-ed: Mac Thornberry. "Washington Won't Solve Our Drought." *USA Today,* August 10, 2011.

The report: Peter Schwartz and Doug Randall. "An Abrupt Climate Change Scenario and Its Implications for United States Security." US Department of Defense, October 2003.

"threat multipliers": John D. Banusiewicz. "Hagel to Address 'Threat Multiplier' of Climate Change." DoD News, October 13, 2014.

Quadrennial Defense Review: Hagel, Quadrennial Defense Review, 34.

Mattis: Andrew Revkin. "Trump's Defense Secretary Cites Climate Change as National Security Challenge." ProPublica, March 14, 2017. Accessed March 20, 2017. https://www.propublica.org/article/trumps-defense-secretary-cites-climate-change-national-security-challenge

Senate floor: John McCain. "Remarks on Climate Stewardship Act of 2007." Office of Senator John McCain press release. January 12, 2007.

Center on Climate Change and National Security: Annie Snider. "Amid Budget Scrutiny, CIA Shutters Climate Center." Greenwire, November 19, 2012. Accessed December 20, 2016. http://www.eenews.net/stories/1059972724

amendment: Ryan Koronowski. "House Votes to Deny Climate Science and Ties Pentagon's Hands on Climate Change." ThinkProgress, May 22, 2014. Accessed March 1, 2017. https://thinkprogress.org/house-votes-to-deny-climate-science-and-ties-pentagons-hands-on-climate-change-6fb577189fb0#.bd9dd1dwq

"When we distract…": W. J. Hennigan. "Climate Change Is Real: Just Ask the Pentagon." *Los Angeles Times,* November 11, 2016.

"mass destruction…": Arshad Mohammed. "Kerry Calls Climate Change 'Weapon of Mass Destruction.'" Reuters, February 16, 2014.

"On what planet…": Aaron Blake. "Gingrich Calls for Kerry to Resign over Climate Change Speech." *Washington Post,* February 18, 2014.

exhaustive study: Colin P. Kelley et al. "Climate Change in the Fertile Crescent and Implications of the Recent Syrian Drought." *Proceedings of the National Academy of Sciences* 112, no. 11 (2015), 3241–46.

Syrian farmer: Ibid., 3246.

"cripple the security…": Bryan Bender. "Chief of US Pacific Forces Calls Climate Biggest Worry." *Boston Globe,* March 9, 2013.

Inhofe: Ryan Koronowski. "Congress: Where the Bible Disproves Science and a Senator Tries to Torpedo an Admiral." ThinkProgress, April 10, 2013. Accessed March 7, 2017. https://thinkprogress.org/congress-where-the-bible-disproves-science-and-a-senator-tries-to-torpedo-an-admiral-73dc1772710

global CO_2 levels: When Kerry and I talked in late 2015, the latest annual CO_2 emissions data showed a long upward trend. In 2015 and 2016, due largely to reduced coal consumption in China and improved energy efficiency in the US, the upward curve flatlined at about thirty-two gigatons per year of CO_2. "IEA Finds CO_2 Emissions Flat for Third Straight Year Even as Global Economy Grew in 2016." International Energy Agency, March 17, 2017. Accessed March 24, 2017. https://

www.iea.org/newsroom/news/2017/march/iea-finds-co2-emissions
-flat-for-third-straight-year-even-as-global-economy-grew.html

Chapter 10

official count: United Nations Data Booklet. "The World's Cities in 2016."
Accessed March 10, 2017. http://www.un.org/en/development/
desa/population/publications/pdf/urbanization/the_worlds_cities
_in_2016_data_booklet.pdf

21 million: National Population Commission, Nigeria. The easiest way to
access its population count is here: https://www.citypopulation
.de/php/nigeria-metrolagos.php.

ten times faster: Walter Leal Filho and Ulisses M. Azeiteiro, eds. *Climate
Change and Health: Improving Resilience and Reducing Risks* (New
York: Springer, 2016), 175.

$1.25 a day: "World Development Indicators 2013." The World Bank,
2013. Accessed March 7, 2017. http://hdr.undp.org/en/content/
population-living-below-125-ppp-day

two million barrels: "OPEC Annual Statistical Bulletin: Organization of
the Petroleum Exporting Countries," 2016. Accessed March 7, 2017.
http://www.opec.org/opec_web/static_files_project/media/
downloads/publications/ASB2016.pdf

Flash flooding: "Nigeria Floods Kill 363 People, Displace 2.1 Million."
Reuters, November 5, 2012.

Eko Atlantic: The development has a glitzy website, which includes a virtual tour of the site: http://www.ekoatlantic.com/

South China Sea: Gordon Lubold. "Pentagon Says China Has Stepped
Up Land Reclamation in South China Sea." *Wall Street Journal,*
August 20, 2015.

Singapore: "Such Quantities of Sand." *The Economist,* February 26, 2015.

Tokyo Bay: Ibid.

satellite data: Alister Doyle. "Coastal Land Expands as Construction Out-
paces Sea-Level Rise." Reuters, August 25, 2016.

Frontline: Robin Urevich. "Chasing the Ghosts of a Corrupt Regime: Gilbert Chagoury, Clinton Donor and Diplomat with a Checkered
Past." *Frontline/World,* January 8, 2010.

economic losses: Stephane Hallegatte et al. "Future Flood Losses in
Major Coastal Cities." *Nature Climate Change 3,* no. 9 (2013), 802–6.

30 million: Daniel Hoornweg and Kevin Pope. "Socioeconomic Pathways
and Regional Distribution of the World's 101 Largest Cities." Global
Cities Institute working paper, 2014 Accessed March 4, 2017. http://

media.wix.com/ugd/672989_62cfa13ec4ba47788f78ad660489 a2fa.pdf

sub-Saharan coastline: Matteo Fagotto. "West Africa Is Being Swallowed by the Sea." *Foreign Policy,* October 21, 2016.

"In West Africa…": Quoted in ibid.

Accra: Ibid.

breeding grounds for sea turtles: Ibid.

"Some of our children…": Ibid.

floating school: "Makoko Floating School/NLE Architects." ArchDaily, March 14, 2013. Accessed March 4, 2017. http://www.archdaily .com/344047/makoko-floating-school-nle-architects

"beacon of hope": Jessica Collins. "Makoko Floating School, Beacon of Hope for the Lagos 'Waterworld.'" *The Guardian,* June 2, 2015.

250,000 used plastic bottles: "The Island in Cancun Built on Recycled Plastic Bottles." BBC News, April 2, 2016.

Seasteading Institute: Kyle Denuccio. "Silicon Valley Is Letting Go of Its Techie Island Fantasies." Wired.com, May 16, 2015. Accessed March 5, 2017. https://www.wired.com/2015/05/silicon-valley-letting -go-techie-island-fantasies/

"serious blow": Cynthia Okoroafor. "Does Makoko Floating School's Collapse Threaten the Whole Slum's Future?" *The Guardian,* June 10, 2016.

300,000 people: Unofficial estimate, provided by Megan Chapman of Justice & Empowerment Initiatives in Lagos.

issued an order: Ben Ezeamalu. "Lagos Slum Dwellers Set for Showdown with Government over Eviction Notice." *Premium Times,* October 12, 2016.

police entered: Paola Totaro and Matthew Ponsford. "Demolitions of Lagos Waterfront Communities Could Leave 300,000 Homeless: Campaigners." Reuters, November 11, 2016.

Otodo Gbame: Laurin-Whitney Gottbrath. "Thousands Displaced as Police Raze Lagos' Otodo Gbame." Aljazeera.com, April 10, 2017. Accessed May 2, 2017. http://www.aljazeera.com/news/2017/04/thousands -displaced-police-raze-lagos-otodo-gbame-170410090717831.html

Chapter 11

Miami Beach developer: Zachery Fagenson. "Sunset Harbour Developer Scott Robins: It's Never the Chef, It's the Business Guy." *Miami New Times,* December 28, 2015. See also Christina Lawrence. "Astute Awakening." *Miami,* October 24, 2012. Accessed March 7, 2017. http://www.modernluxury.com/miami/articles/astute-awakening

made a fortune: Richard Bradley. "Philip Levine's Second Wave." *Worth,* October 7, 2014.

$100 million in bonds: "Moody's Assigns Negative Outlook to Miami Beach, Florida's Stormwater Revenue Bonds." Moody's Investors Service, July 10, 2015. Accessed February 4, 2017. https://www .moodys.com/research/Moodys-assigns-negative-outlook-to-Miami -Beach-FLs-Stormwater-Revenue--PR_329912

Rolling Stone: Jeff Goodell. "Goodbye, Miami." *Rolling Stone,* June 20, 2013. See also Suzanne Goldenberg. "US East Coast Cities Face Frequent Flooding Due to Climate Change." *The Guardian,* October 8, 2014. See also Joel Achenbach. "Is Miami Drowning?" *Washington Post,* July 16, 2014.

Chicago: The engineering and political complexities of raising Chicago are covered in detail in Harold L. Platt. *Shock Cities: The Environmental Transformation and Reform of Manchester and Chicago* (Chicago: University of Chicago Press, 2005), 118–133.

cholera epidemic: David Young. "Raising the Chicago Streets Out of the Mud." *Chicago Tribune,* November 15, 2015.

fecal levels: Jenny Staletovich. "Miami Beach King Tides Flush Human Waste into Bay, Study Finds." *Miami Herald,* May 16, 2016.

"in order to sell ads": Fred Grimm. "The Stink Beach Mayor Smells Isn't a Conspiracy, It's Fecal Runoff." *Miami Herald,* June 9, 2016.

a liar: Personal communication between the author and an off-the-record source.

"a hit job": Grimm, "The Stink Beach Mayor Smells Isn't a Conspiracy, It's Fecal Runoff."

"recklessly and incorrectly...": Letter from Raul Aguila, Miami Beach city attorney, to Aminda Marqués Gonzalez, executive editor of *Miami Herald,* May 25, 2016. Accessed March 15, 2017. http://www .miamiherald.com/latest-news/article82543332.ece/binary/ Letter%20To%20Aminda%20Gonzalez.pdf

$1.6 billion on repairs: Patricia Mazzei. "Federal Judge Signs Agreement for $1.6 Billion in Miami-Dade Sewer Repairs." *Miami Herald,* April 15, 2014.

86,000 in-ground systems: Linda Young. "Florida Waters: 'Fountains of Youth' or 'Fountains of Yuk'?" Report for the Florida Clean Water Network, February 13, 2015. Accessed March 12, 2017. http:// floridacleanwaternetwork.org/florida-waters-fountains-of-youth -or-fountains-of-yuk/

Florida Department of Health: Personal communication with the author.

40 percent: Young, "Florida Waters: 'Fountains of Youth' or 'Fountains of Yuk'?"

green glop: Craig Pittman. "Toxic Algae Bloom Crisis Hits Florida, Drives Away Tourists." *Tampa Bay Times,* July 1, 2016.

viral tracers: John H. Paul et al. "Viral Tracer Studies Indicate Contamination of Marine Waters by Sewage Disposal Practices in Key Largo, Florida." *Applied and Environmental Microbiology* 61, no. 6 (1995), 2230.

Haiti: Jonathan M. Katz. "UN Admits Role in Cholera Epidemic in Haiti." *New York Times,* August 17, 2016.

Mount Trashmore: Sean Rowe. "Our Garbage, Ourselves." *Miami New Times,* January 25, 1996.

remains to float out: Lydia O'Connor. "Even the Dead Have Been Displaced by Louisiana Flooding." *Huffington Post,* August 19, 2016. Accessed March 12, 2017. http://www.huffingtonpost.com/entry/louisiana-flooding-caskets_us_57b5e6d7e4b034dc73262ee2

cemeteries are on low-lying ground: I calculated the elevations of cemeteries mentioned here with Google Maps' Elevation service. I accessed the service through a Web app developed by Florida International University. Accessed March 12, 2017. http://citizeneyes.org/app/

Key West Cemetery: Personal communication with Russell Brittain, Sexton at Key West Cemetery.

elevated twenty feet: Personal communication with Michael Waldron, Florida Power & Light spokesperson. April 2013. Also mentioned in Christina Nunez, "As Seas Rise, Are Coastal Nuclear Plants Ready?" *National Geographic,* December 16, 2015.

peak storm surge: Ed Rappaport. "Preliminary Report: Hurricane Andrew, 16–28 August, 1992." National Hurricane Center. Accessed March 12, 2017. http://www.nhc.noaa.gov/1992andrew.html

leaky canals: Jenny Staletovich. "Evidence of Salt Plume Under Turkey Point Nuclear Plant Goes Back Years." *Miami Herald,* April 21, 2016.

inundation maps: I used Climate Central's Surging Seas risk finder. Accessed February 14, 2017. http://sealevel.climatecentral.org

$20-billion plan: Susan Salisbury. "FPL's Turkey Point Cost Estimate Rises to Top Range of $20 Billion." *Palm Beach Post,* June 27, 2015.

environmental review: "FPL Gets Environmental Approval for Two More Reactors at Turkey Point." *Miami Herald,* November 3, 2016.

Chapter 12

Snowmass: "Workshop on Critical Issues in Climate Change." Energy Modeling Forum. July 25–August 3, 2006. Details of event reconstructed from interviews with many participants, including Lowell Wood.

"multiplanet civilization": Quoted in Ross Andersen. "Exodus." Aeon, September 30, 2014. Accessed March 12, 2017. https://aeon.co/essays/elon-musk-puts-his-case-for-a-multi-planet-civilisation

research project: Henry Fountain. "White House Urges Research on Geoengineering to Combat Climate Change." *New York Times,* January 10, 2017.

Davos: The Global Risks Report 2017. World Economic Forum, Geneva, 43. Accessed March 12, 2017. http://www3.weforum.org/docs/GRR17_Report_web.pdf

Antarctica: Chris Mooney. "This Mind-Boggling Study Shows Just How Massive Sea-Level Rise Really Is." *Washington Post,* March 10, 2016.

$2 billion a year: David Keith. *A Case for Climate Engineering* (Cambridge, MA: MIT Press, 2013), 43.

subsidies: Coming up with an accurate number for fossil fuel subsidies is difficult, in part because it depends on how you define a subsidy. My $1 trillion estimate comes from Oil Change International's "Fossil Fuel Subsidies: Overview." Accessed March 12, 2017. http://priceofoil.org/fossil-fuel-subsidies/
If "externalities" such as the health effects of air pollution, the environmental damages caused by drilling and mining, and the impacts of climate change are included, the cost of fossil fuel subsidies estimated by the International Monetary Fund rises to more than $5 trillion annually. Accessed March 12, 2017. http://www.imf.org/external/pubs/ft/survey/so/2015/NEW070215A.htm

6.5 million people: Stanley Reed. "Study Links 6.5 Million Deaths Each Year to Air Pollution." *New York Times,* June 26, 2016.

Keith estimated: Keith, *A Case for Climate Engineering,* 69.

cancer clusters: Dan Fagin. *Toms River: A Story of Science and Salvation* (New York: Bantam, 2013), 332.

a report: "Under Water: How Sea-Level Rise Threatens the Tri-State Region." A report of the Fourth Regional Plan. Regional Plan Association, December 2016, 18. Accessed March 2, 2017. http://library.rpa.org/pdf/RPA-Under-Water-How-Sea-Level-Rise-Threatens-the-Tri-State-Region.pdf

nine-foot storm surge: "Barnegat Bay Storm Surge Elevations During Hurricane Sandy." The Richard Stockton College of New Jersey, October 29, 2014, 14. Accessed March 2, 2017. http://www.nj.gov/dep/shoreprotection/docs/ibsp-barnegat-bay-storm-surge-elevations-during-sandy.pdf

10,000 homes: Jill P. Capuzzo. "Not Your Mother's Jersey Shore." *New York Times,* June 16, 2017.

all but 60: Ibid.

one document: "Resilience + the Beach: A Regional Strategy and Pilot Projects for the Jersey Shore." Jury brief by Rutgers University, Saski, and ARUP for Rebuild by Design Competition, March 2014, 19. Accessed March 2, 2017. http://www.rebuildbydesign.org/data/files/670.pdf

stronger dune: Karen Wall. "Protection for Toms River: Long-Awaited Army Corps Dune Project Goes Out to Bid." Toms River Patch, September 29, 2016.

shortfall: Personal communication with Mayor Tom Kelaher's office, January 2017.

building codes: Gregory Kyriakakis. "Toms River Continues Aim to Relax Construction Rules for Sandy-Damaged Homes." Toms River Patch, May 8, 2013.

FEMA: Leslie Kaufman. "Sandy's Lessons Lost: Jersey Shore Rebuilds in Sea's Inevitable Path." Inside Climate News, October 26, 2016. Accessed February 20, 2017. https://insideclimatenews.org/news/25102016/hurricane-sandy-new-jersey-shore-rebuild-climate-change-rising-sea-chris-christie

$4.6 billion: State of New Jersey, Office of the State Comptroller. NJ Sandy Transparency funds tracker. Accessed March 9, 2017. http://nj.gov/comptroller/sandytransparency/funds/tracker/
The total projected spending for Sandy recovery in New Jersey is $9 billion, but so far, only $4.6 billion has been spent. According to Lisa Ryan, director of strategic communication for Sandy recovery, New Jersey Department of Community Affairs, 95 percent of that $4.6 billion has been federal funds. Personal communication, March 21, 2017.

$300 million in federal recovery funds: Personal communication with Mayor Tom Kelaher's office, January 2017.

$30 million: Ibid.

610 properties: "NY Rising 2012–2016: Fourth Anniversary Report." Governor's Office of Storm Recovery, 8. Accessed March 9, 2017. https://stormrecovery.ny.gov/sites/default/files/crp/community/documents/10292016_GOSR4thAnniversary.pdf

resettle twenty-three families: Coral Davenport and Campbell Robertson. "Resettling the First American 'Climate Refugees.'" *New York Times,* May 2, 2016.

Coastal Master Plan: "Louisiana's Comprehensive Master Plan for a Sustainable Coast." Coastal Protection and Restoration Authority of Louisiana, 2017, 145. Accessed March 9, 2017. http://coastal.la.gov/wp-content/uploads/2016/08/2017-MP-Book_Single_Combined_01.05.2017.pdf

Newtok: "Alaska Seeks Federal Money to Move a Village Threatened by Climate Change." The Associated Press, October 3, 2015.

Flavelle: Christopher Flavelle. "The Toughest Question in Climate Change: Who Gets Saved?" Bloomberg View, August 29, 2016. Accessed March 8, 2017. https://www.bloomberg.com/view/articles/2016-08-29/the -toughest-question-in-climate-change-who-gets-saved

Three Gorges: Jim Yardley. "Chinese Dam Projects Criticized for Their Human Costs." *New York Times*, November 19, 2007.

Nijmegen: Mathieu Schouten. "Partnering a River." *My Liveable City*, January–March 2016, 68–73. See also the Room for the River website: https://www.ruimtevoorderivier.nl/english/

Oakwood Beach: Jada Yuan. "Last Stand on Oakwood Beach." *New York*, March 3, 2013.

common law: Peter J. Byrne. "The Cathedral Engulfed: Sea-Level Rise, Property Rights, and Time." *Louisiana Law Review* 73, no. 12 (2012), 69–118.

Summer Haven: Sue Bjorkman. "Good Ole Summer Haven Time." Old CityLife.com, September 29, 2016. http://www.oldcitylife.com/ features/good-ole-summer-haven-time/

road washed out: Ken Lewis. "Great Location, Lovely View, but There's No Road." *Florida Times-Union*, August 16, 2005.

annual maintenance costs: Personal communication with St. Johns County attorney Patrick F. McCormack, January 26, 2017.

$950,000 in federal funds: Ibid.

sued the county: *Robert and Linnie Jordan et al. v. St. Johns County*, case no. CA05-694 (Florida Seventh Judicial Circuit, May 21, 2009).

after Hurricane Katrina: Edward P. Richards. "The Hurricane Katrina Levee Breach Litigation: Getting the First Geoengineering Liability Case Right." *University of Pennsylvania Law Review*, 2012, vol. 160, issue 1, article 13. Accessed March 12, 2017. http://scholarship.law .upenn.edu/penn_law_review_online/vol160/iss1/13

"ravages of the ocean": Cited in Thomas Ruppert and Carly Grimm. "Drowning in Place: Local Government Costs and Liabilities for Flooding Due to Sea-Level Rise." *Florida Bar Journal*, November 2013, vol. 87, no. 9, 29–33.

district court: *Robert and Linnie Jordan et al. v. St. Johns County*, case no. 5D09-2183 (Florida Fifth District Court of Appeal, May 20, 2011).

voice on my radio: Ryan Kailath. "Louisiana Tries New Defense Against Floods: Move People to Higher Ground." NPR, January 29, 2017. Accessed March 20, 2017. www.npr.org/2017/01/29/512271883/ louisiana-tries-new-defense-against-floods-move-people-to-higher -ground

Epilogue

Denmark: Walter Alvarez. *T. Rex and the Crater of Doom* (Princeton: Princeton University Press, 1997), 70.

"It was clear…": Ibid., 71.

"During the time…": Ibid.

100 percent renewables: "Renewable Energy." Danish Ministry of Energy, Utilities, and Climate. Accessed March 2, 2017. http://old.efkm.dk/en/climate-energy-and-building-policy/denmark/energy-supply-and-efficiency/renewable-energy

TNT: Elizabeth Kolbert. *The Sixth Extinction: An Unnatural History* (New York: Henry Holt and Company, 2014), 75.

Louisiana: Coastal Protection and Restoration Authority of Louisiana, 13.

"managed realignment": Nigel Pontee. "Factors Influencing the Long-Term Sustainability of Managed Realignment." *Managed Realignment: A Viable Long-Term Coastal Management Strategy?* (New York: Springer Briefs in Environmental Science, 2014), 95–107.

Selected Bibliography

Adams, Mark. *Meet Me in Atlantis: My Obsessive Quest to Find the Sunken City*. New York: Penguin, 2015.

Alley, Richard B. *The Two-Mile Time Machine: Ice Cores, Abrupt Climate Change, and Our Future*. Princeton: Princeton University Press, 2014.

Allman, T. D. *Finding Florida: The True History of the Sunshine State*. New York: Grove/Atlantic, 2013.

———. *Miami: City of the Future*. Gainesville: University Press of Florida, 2013.

Alvarez, Walter. *T. Rex and the Crater of Doom*. Princeton: Princeton University Press, 1997.

Armstrong, Karen. *A Short History of Myth*. Edinburgh: Canongate, 2005.

Ballard, J. G. *The Drowned World: A Novel*. New York: W. W. Norton and Company, 2012.

Barker, Robert, and Richard Coutts. *Aquatecture: Buildings and Cities Designed to Live and Work with Water*. Newcastle upon Tyne: RIBA Publishing, 2016.

Berendt, John. *The City of Falling Angels*. New York: Penguin, 2006.

Brodsky, Joseph. *Watermark: An Essay on Venice*. London: Penguin UK, 2013.

Burgis, Tom. *The Looting Machine: Warlords, Oligarchs, Corporations, Smugglers, and the Theft of Africa's Wealth*. New York: PublicAffairs, 2015.

Byron, George Gordon. *Lord Byron: The Major Works*. Oxford: Oxford University Press, 2008.

Carson, Rachel. *The Sea Around Us.* New York: Oxford University Press, 1951.

Clark, Nancy, and Kai-Uwe Bergmann. *Miami Resiliency Studio.* Gainesville: University of Florida, 2015.

Didion, Joan. *Miami.* New York: Simon and Schuster, 1987.

Dobbs, David. *Reef Madness: Charles Darwin, Alexander Agassiz, and the Meaning of Coral.* New York: Pantheon, 2009.

Englander, John. *High Tide on Main Street: Rising Sea Level and the Coming Coastal Crisis.* The Science Bookshelf, 2012.

Fagan, Brian. *The Attacking Ocean: The Past, Present, and Future of Rising Sea Levels.* New York: Bloomsbury Publishing, 2014.

Fagin, Dan. *Toms River: A Story of Science and Salvation.* New York: Bantam, 2013.

Finkel, Irving. *The Ark Before Noah: Decoding the Story of the Flood.* New York: Nan A. Talese, 2014.

Fisher, Jerry M. *The Pacesetter: The Untold Story of Carl G. Fisher.* Victoria, BC: FriesenPress, 2014.

Fletcher, Caroline, and Jane Da Mosto. *The Science of Saving Venice.* Turin, Italy: Umberto Allemandi and Co., 2004.

Gillis, John R. *The Human Shore: Seacoasts in History.* Chicago: University of Chicago Press, 2012.

Gould, Stephen Jay. *Leonardo's Mountain of Clams and the Diet of Worms.* New York: Harmony Books, 1998.

Grunwald, Michael. *The Swamp: The Everglades, Florida, and the Politics of Paradise.* New York: Simon and Schuster, 2007.

Harari, Yuval Noah. *Sapiens: A Brief History of Humankind.* New York: HarperCollins, 2015.

Hazen, Robert M. *The Story of Earth: The First 4.5 Billion Years, from Stardust to Living Planet.* New York: Penguin, 2013.

Hine, Albert C. *Geologic History of Florida: Major Events That Formed the Sunshine State.* Gainesville: University Press of Florida, 2013.

Hobbs, Carl H. *The Beach Book: Science of the Shore.* New York: Columbia University Press, 2012.

Keith, David. *A Case for Climate Engineering.* Cambridge, MA: MIT Press, 2013.

Keith, Vanessa, and Studioteka. *2100: A Dystopian Utopia: The City After Climate Change.* New York: Terreform, 2017.

Kolbert, Elizabeth. *The Sixth Extinction: An Unnatural History.* New York: Henry Holt and Company, 2014.

Leary, Jim. *The Remembered Land: Surviving Sea-Level Rise After the Last Ice Age.* New York: Bloomsbury Publishing, 2015.

Lencek, Lena, and Gideon Bosker. *The Beach: The History of Paradise on Earth.* New York: Penguin, 1999.

Macaulay, Rose. *Pleasure of Ruins.* London: Thames and Hudson, 1953.

Madden, Thomas F. *Venice: A New History.* New York: Viking, 2012.

McGrath, Campbell. *Florida Poems.* New York: HarperCollins, 2003.

Meltzer, David J. *First Peoples in a New World: Colonizing Ice Age America.* Berkeley: University of California Press, 2009.

Metz, Tracy, and Maartje van den Heuvel. *Sweet and Salt: Water and the Dutch.* Rotterdam: NAi Publishers, 2012.

Mitchell, Stephen. *Gilgamesh.* New York: Free Press, 2004.

Montgomery, David R. *The Rocks Don't Lie: A Geologist Investigates Noah's Flood.* New York: W. W. Norton and Company, 2012.

Morton, Oliver. *The Planet Remade: How Geoengineering Could Change the World.* Princeton: Princeton University Press, 2015.

Ogden, Laura A. *Swamplife: People, Gators, and Mangroves Entangled in the Everglades.* Minneapolis: University of Minnesota Press, 2011.

Oka Doner, Michele, and Mitchell Wolfson Jr. *Miami Beach: Blueprint of an Eden: Lives Seen Through the Prism of Family and Place.* New York: HarperCollins, 2007.

Oshima, Ken Tadashi (ed.). *Between Land and Sea: Kiyonori Kikutake.* Zurich: Lars Müller, 2015.

Parks, Arva Moore. *The Forgotten Frontier: Florida Through the Lens of Ralph Middleton Munroe.* Miami: Centennial Press, 2004.

———. *George Merrick, Son of the South Wind: Visionary Creator of Coral Gables.* Gainesville: University Press of Florida, 2016.

Pérez, Jorge. *Powerhouse Principles: The Ultimate Blueprint for Real Estate Success in an Ever-Changing Market.* New York: Penguin, 2008.

Pittman, Craig. *Oh, Florida!: How America's Weirdest State Influences the Rest of the Country.* New York: St. Martin's Press, 2016.

Platt, Harold L. *Shock Cities: The Environmental Transformation and Reform of Manchester and Chicago.* Chicago: University of Chicago Press, 2005.

Purdy, Jedediah. *After Nature: A Politics for the Anthropocene.* Cambridge, MA: Harvard University Press, 2015.

Redford, Polly. *Billion-Dollar Sandbar: A Biography of Miami Beach.* New York: Dutton, 1970.

Rudiak-Gould, Peter. *Surviving Paradise: One Year on a Disappearing Island.* New York: Sterling, 2009.

Ryan, William, and Walter Pitman. *Noah's Flood: The New Scientific Discoveries About the Event That Changed History.* New York: Simon and Schuster, 2000.

Schober, Theresa M. *Art Calusa: Reflections on Representation.* Fort Myers: Lee Trust for Historic Preservation, 2013.

Settis, Salvatore. *If Venice Dies.* New York: New Vessel Press, 2016.

Shearer, Christine. *Kivalina: A Climate Change Story.* Chicago: Haymarket Books, 2011.

Shepard, Francis P., and Harold R. Wanless. *Our Changing Coastlines*. New York: McGraw-Hill, 1971.

Shepard, Jim. *You Think That's Bad: Stories*. New York: Knopf, 2011.

Shorto, Russell. *Amsterdam: A History of the World's Most Liberal City*. New York: Vintage, 2013.

Sobel, Adam. *Storm Surge: Hurricane Sandy, Our Changing Climate, and Extreme Weather of the Past and Future*. New York: HarperCollins, 2014.

Sullivan, Walter. *Assault on the Unknown: The International Geophysical Year*. New York: McGraw-Hill, 1961.

Walker, Gabrielle. *Antarctica: An Intimate Portrait of a Mysterious Continent*. Boston: Houghton Mifflin Harcourt, 2013.

Ward, Peter D. *The Flooded Earth: Our Future in a World Without Ice Caps*. New York: Basic Books, 2010.

Williams, Joy. *The Florida Keys: A History and Guide*. New York: Random House, 2010.

Index

About the Author

Jeff Goodell is a contributing editor at *Rolling Stone* and the author of five books, most recently *How to Cool the Planet: Geoengineering and the Audacious Quest to Fix Earth's Climate,* which won the 2011 Grantham Prize Award of Special Merit. Goodell's previous books include *Sunnyvale,* a memoir about growing up in Silicon Valley, which was a *New York Times* Notable Book, and *Big Coal: The Dirty Secret Behind America's Energy Future.*

ONE OF THE BEST BOOKS OF THE YEAR
New York Times · *Washington Post* · *Booklist*

"Immersive…a powerful reminder that we can bury our heads in the
sand about climate change for only so long before the sand itself disappears."

—Jennifer Senior, *New York Times* (Top Ten Books of the Year)

A cross the globe, scientists and civilians are noticing rapidly rising sea levels and increasingly higher tides that push more water directly into the places we live—from our most vibrant, historic cities to our last remaining traditional coastal villages. By this century's end, hundreds of millions of people will be retreating from the world's shores, a harrowing crisis of social, environmental, and fiscal measures. Yet despite international efforts and tireless research, there is no permanent solution—no barriers to erect or walls to build—that will protect us in the end from the drowning of the world as we know it.

The Water Will Come is the definitive examination of this coming catastrophe, why and how it will happen, and what the endgame will mean for the way we live. Traveling across twelve countries, Jeff Goodell reports from the front lines of the climate change epidemic, revealing to us the water world into which our planet is quickly transforming.

"Read this book for a reminder of the stakes—right now, today—and why we have to work harder, faster, to address the climate challenge."

—John F. Kerry

"*The Water Will Come* is absolutely brilliant scientific journalism, and certainly is a must-read for all of the world's citizens."

—*Forbes*

JEFF GOODELL
is a contributing editor at *Rolling Stone* and the author of five books, including *How to Cool the Planet*, which won the 2011 Grantham Prize Award of Special Merit; *Sunnyvale*, a memoir of growing up in Silicon Valley; and *Big Coal: The Dirty Secret Behind America's Energy Future*.

 @jeffgoodell

littlebrown.com
Twitter @littlebrown
Facebook.com/littlebrownandcompany

ALSO AVAILABLE IN AUDIO AND EBOOK EDITIONS
COVER DESIGN BY NEIL ALEXANDER HEACOX
COVER PHOTOGRAPH BY THOMAS JACKSON / GETTY IMAGES
AUTHOR PHOTOGRAPH BY PERNILLE AEGIDIUS
COVER © 2018 HACHETTE BOOK GROUP, INC.
PRINTED IN THE U.S.A.